Measuring Electronics and Sensors

Herbert Bernstein

Measuring Electronics and Sensors

Basics of Measurement Technology, Sensors, Analog and Digital Signal Processing

 Springer

Herbert Bernstein
Technikerschule, Elektroinnung und IHK in München
München, Germany

This book is a translation of the original German edition „Messelektronik und Sensoren" by Bernstein, Herbert, published by Springer Fachmedien Wiesbaden GmbH in 2014. The translation was done with the help of artificial intelligence (machine translation by the service DeepL.com). A subsequent human revision was done primarily in terms of content, so that the book will read stylistically differently from a conventional translation. Springer Nature works continuously to further the development of tools for the production of books and on the related technologies to support the authors.

ISBN 978-3-658-35066-6   ISBN 978-3-658-35067-3   (eBook)
https://doi.org/10.1007/978-3-658-35067-3

This Springer imprint is published by the registered company Springer Fachmedien Wiesbaden GmbH part of Springer Nature.
The registered company address is: Abraham-Lincoln-Str. 46, 65189 Wiesbaden, Germany

Preface

Rational operation management in process plants is no longer conceivable without modern measuring technology. A prerequisite for the use of control devices, controllers, process computers and PC systems is a continuous or quasi-continuous and exact recording of measured values and the representation of the measured values in the form of electrical, hydraulic or pneumatic signals.

The metrological recording and evaluation of the environment have always been the prerequisite for the engineer and physicist's work. The scientific and technical tasks that they face today present them with a very large number of interrelated problems. Their solution requires a holistic approach to which the methods of sensor technology and measurement data acquisition and processing can make an important contribution. For the technician and master craftsman in the laboratory, in operation and in plant engineering, the practical application of the numerous sensors and the knowledge of measurement electronics or PC measurement technology are indispensable today.

The book gives an insight into today's operational measurement technology including analytical technology, without claiming to be complete. For the student, the book is an introduction in addition to the relevant textbooks and manuals. It gives the engineer in the profession a quick overview of measurement methods and instruments that are not familiar to him.

In measurement technology, a PC system has become an irreplaceable aid. The PC automatically carries out measurements, analyses measurement data and makes them available as graphics or files for further processing. The PC can control entire plants by switching relays, valves and motors. Of course, it is also indispensable in control engineering. In the past, a control circuit had to be built up using numerous discrete components or very expensive process computers. Today, it can be made more flexible and effective by using a PC system because hardware components can be replaced by appropriate software.

The necessary combination of analog and digital circuit technology often results in problems that can only be solved with the appropriate experience in both areas. For this reason, this book not only presents the components of measurement technology in a transparent manner, but also provides a detailed description of the analog components required for the design of measurement and control systems.

The experience and previous knowledge of PC users or those who want to become PC users are of very different nature. On the one hand, there are the "PC freaks" who have high level of knowledge about programming and the inner workings of the PC, but are not so familiar with analog technology that they could develop appropriate systems. On the other hand, the PC is used in a wide variety of areas, whether in electrical engineering/electronics, mechanical engineering or process engineering. Here, the PC system is rather regarded as an aid for the actual problem solution and accordingly the PC knowledge is usually not profound enough to be able to judge measuring systems regarding their suitability for the PC or to develop themselves.

The theoretical principles and the measuring methods are as much a part of the course as the description of systems, devices and measuring equipment. By specifying measuring ranges and error limits, additional indications for use are given, whereby the values given are to be regarded as minimum values due to the constant technical development. Numerous tables round off the book into a handy working document.

I would like to thank my wife Brigitte for preparing the drawings and for correcting the manuscript.

Munich, Germany   Herbert Bernstein
Summer 2020

Contents

Introduction to Sensor Technology and Electronic Measurement Technology

1

Summary

At the beginning of every electrical measurement of non-electrical quantities are the sensors. They use certain physical effects by which the measured variable is converted into an electrical variable. In most cases, these effects are not discoveries of our time, but in some cases have been known for decades or centuries. They have merely been "bred up" in the direction of simple applicability within the scope of today's possibilities of electronics, PC measurement technology, and microcontrollers in front-end systems.

The operation of many sensors has been known for a long time. For example, photocells and thermocouples were already used in the 1930s for light and temperature measurement. However, the spread of sensors only really began in the 1970s, when temperature- and pressure-sensitive semiconductor components appeared and the word "sensor", which is used in English, also became established in the German language.

In all considerations of "sensors", a distinction shall be made between "sensor elements" and "sensor systems":

- Sensor elements are actually probes that convert a physical quantity into an electrical signal. Often they are manufactured by a component manufacturer and cannot be used directly in rough industrial environments in their delivered form, because they do not yet have the right housing and the necessary signal pre-processing. In many cases, other companies take over this production step.
- Sensor systems, on the other hand, contain not only the sensor elements built into a practical housing, but also part of the electronics for its operation and the processing of the signal obtained—such as pre-amplification, linearization, temperature compensation, and adaptation to further signal processing, in some cases even digitization for direct connection to a computer.

© Springer Fachmedien Wiesbaden GmbH, part of Springer Nature 2022
H. Bernstein, *Measuring Electronics and Sensors*,
https://doi.org/10.1007/978-3-658-35067-3_1

The developer of electronic regulation and control systems shall decide in each individual case whether to use a sensor element or a sensor system. The considerations will primarily be based on the application—for example, a moisture measurement in a tumble dryer will be carried out differently than in a chemical process. The very wide range of sensor elements and systems available enables the skilled developer to find the optimum solution in every case.

1.1   Sensor Types

Sensors are available for a wide range of measured variables. Their variety is almost unmanageable. The first overview is given in the following compilation, in which the different types are listed according to their share of sales. These figures are based on investigations of the AMA Association for Sensor and Measurement Technology eV (AMA Fachverbandes für Sensorik e. V. E). They are approximate estimates, the exact values are sometimes difficult to determine in practice.

Level sensors	20%
Pressure, differential pressure	20%
Temperature	15%
Flow rate	10%
Speed	10%
Force, weight	8%
Acceleration	2%
Humidity	1%
Electrochemistry	1%
Gas analysis	1%
Other	12%
Total	100%

The highest growth rates, in some cases significantly more than 10% per year, are seen in distance, acceleration, speed, and rotational speed sensors as well as optical sensors and various biosensors. The use of the sensors can be divided roughly as follows:

Automotive	49%
Budget	27%
Measurement, control, regulation	14%
Data and communications engineering	5%
Consumer electronics	5%

The ideal sensor converts only one single physical, chemical, or geometrical quantity into an electrical signal in a highly linear way and is insensitive to other quantities. In practice, this is often difficult to achieve; in particular, the temperature influences the measured value for many types. Here, appropriate compensation is necessary.

Whoever is responsible for the selection or development of sensors has to make decisions in the field of sensor physics, sensor technology, sensor production, sensor types, and the very extensive sensor market. About 100 basic physical effects can be used to build sensors. It is important for the developer to "cultivate" the physical effect, that is, to strengthen desired interactions and to attenuate or suppress disturbing ones. The diversity of the partly contradictory requirements forces a structured approach to the conversion of the physical effects into products or processes suitable for industrial use.

This is emphatically proven by a study on sensors in mechanical engineering presented as early as 1986 by the Battelle Institute, which examined the question "Why do sensor projects fail?" The most important of the reasons given were environmental conditions (30%), sensor reliability (27%), and sensor costs (27%). This means that all specific requirements must be considered in their entirety. The sensors must be treated as systems (micro- and subsystems). Proven tools of systems engineering are then available for this purpose. Important elements of this approach consist in the clear specification of

- Sensor environment (environment)
- Functional requirements
- Operational requirements
- Interfaces
- Safety requirements

The following checklist covering ten subject areas can provide help in creating specifications for sensors:

- Functional requirements:
  - Basic physical principle
  - Intermediate sizes
  - Accuracy
  - Resolution
  - Dynamics
  - Compensation of disturbance variables
- Environmental requirements:
  - Climatic requirements
    Temperature
    Pressure
    Humidity
  - Mechanical requirements
    Vibration
    Shock
    Acceleration
  - Electromagnetic requirements
    Resistance to induced disturbances

        Resistance to irradiated interference
        Reduction of the derived interference
        Reduction of radiated interference
- Interface requirements:
  - Electrical interface (operating voltage):
    - Voltage
    - Performance
  - Electrical interfaces (signals):
  - Analog interfaces:
    - Voltage
    - Electricity
    - Frequency
    - Impedance
    - Digital interfaces (parallel):
      - Bit number
      - Protocol
      - Bitrate
      - Signal level
    - Digital interfaces (serial):
      - Baud rate
      - Protocol
      - Configuration
      - Bus agreements (collision avoidance)
    - Mechanical interfaces:
      - Housing (protection class according to DIN)
      - Mounting
      - Connectors
- Signal processing:
  - Analog signal processing:
    - Reinforcement
    - Adjustment
    - Impedance
  - Digital signal processing:
    - Resolution (Format)
    - Sampling rate
    - Data reduction
    - Temporary storage
    - Algorithms
- Safety requirements:
  - Electrical safety
  - System security (fail-safe)

- Reliability:
  - MTBF (mean time between failures)
  - Redundancy
- Calibration and test:
  - Laws, standards, guidelines
- Profitability:
  - Technology
  - The optimal number of pieces

Once the measuring tasks and the associated boundary conditions have been specified, the search for a suitable sensor begins.

- Based on the physical interactions, about 2000 different basic sensor methods are known. These in turn have been converted into about 100,000 available products to date. The joy about this variety is extremely by confusing. Added to this is the fact that a large proportion of the existing sensors on the market are not so easy to obtain.
- Standard sensors are generally inexpensive and robust, but the specifications in the datasheet must be followed exactly. Due to contractual obligations of manufacturer and customer, these types are often not directly available. It may therefore be worthwhile to use spare part sensors from the automotive or household appliance sector.
- Expensive, but available in a wide variety, are sensors from the catalogs of numerous manufacturers and distributors. Information about products or suppliers in systematic order can be found in the major trade fairs such as Interkama (Düsseldorf), Sensor und SPS (Nurenberg), Messcomp (Wiesbaden), Electronica (Munich), and Industriemesse (Hanover).
- The third group of sensors contains special sensors, which are developed or adapted to customer specifications and are usually very expensive. The decision to develop a new special sensor must therefore be very carefully considered.
- First of all, it shall be checked whether a standard catalog sensor with the desired properties cannot be adapted. In addition, development shall t be delegated to experts with specific sensor development experience in order to prevent a supposedly small sensor problem from escalating into a long ordeal.

1.1.1 Standard Sensors

Before a detailed description and application of the standard sensors, Table 1.1 provides a brief overview of the measured value recording and the sensor types. Table 1.1 lists the most important physical, chemical, and geometrical quantities and suitable sensor principles. A distinction can be made between the direct measurement of a non-electrical quantity (temperature, optical radiation, magnetic field, etc.) and indirect measurement via an

Table 1.1 Comparison of the main physical, chemical, and geometrical parameters and the corresponding sensor types

Physical, chemical, or geometrical quantity	Sensor type or measuring principle
Pressure	Strain gauge, silicon membrane, surface wave resonator
Force	Strain gauge, silicon membrane, piezo transducer, surface wave resonator
Temperature	NTC, PTC, thermocouple, Pt and Ni resistance thermometer, quartz, Si elements
Luminous intensity, optical radiation	Photoresistor, photodiode, phototransistor, pyroelectric radiation sensors
Volume, sound	Electrodynamic or capacitive microphones
Magnetic field	Hall generator, field plate, a magnetoresistive sensor
Humidity	Metal oxide, semiconductor, metal layer
Gas mixture	Metal oxides, ion-sensitive FET
Radiation, particles	Geiger-Müller counter tube
pH value	Ag/AgCI electrode, ion-sensitive FET
Position, path, length	Inductive or capacitive proximity switch, optical scanning, ultrasonic
Speed	Capacitive or inductive motion detector, optical method, radar
Acceleration	Strain gauges, piezoelectric, capacitive, inductive
Movement, approach	Radar, motion detectors, displacement, optical and inductive sensors
Fill level	Float sensing, capacitive or optical measurement, ultrasonic, heat conduction
Flow rate	Differential pressure according to Venturi, anemometric, magnetic-inductive, Woltmann meter, Coriolis principle, thermal
Speed	Optical and inductive methods
Angular position	Incremental or absolute measurement using optical, capacitive, or inductive sensors
Torque	Strain gauges, magnetostrictive sensors

intermediate quantity (displacement, flow, velocity, etc.). However, it is not always possible to make this distinction consistently.

Sensors can also be divided into passive and active ones. In the former, for example, a resistance, a capacity, or an inductance changes, while the latter emit a voltage or a current as an output signal. With increasing installation of evaluation or adaptation circuits in the same housing or integration of sensor elements and electronics on one chip, however, this division is becoming less and less important.

In today's motor vehicles you can find up to 50 different sensors—which do not always do useful work. Due to future legal regulations and increasing comfort, this number is likely to rise to well over 100. The harsh operating conditions, such as a temperature range of −40 to +150 °C, rapid temperature changes, vibrations, and aggressive atmospheres, require quality standards that, due to the frequency of this stress during the service life of a car, may well approach those of military and aerospace technology. Many new or improved technologies have been developed here in recent years. For example, in a motor vehicle today there are no longer any silicon pressure sensors, but pressure boxes with diaphragm position detection by a Hall generator. Wheel speed sensors are located in the anti-lock braking system (ABS), which not only withstand very high working temperatures and aggressive environmental conditions but also generate usable signals even when the wheel is turning very slowly.

The application possibilities for sensors in automotive electronics are very extensive, especially in the wide field of engine electronics. In the harsh environment of the engine compartment at temperatures between −40 and +150 °C, not only the sensors but also the control units shall work perfectly and be resistant to the effects of moisture, aggressive media such as salts and oils, and to mechanical shocks. Applications are, for example, the alternator regulator in the alternator, the oil level control, and above all the ignition switchgear.

For a sensor, the approval for automotive use is a high qualitative hurdle. But once this has been overcome, low-cost products are available for other areas of application, although they usually require some "refinement", for example in terms of linearity or measuring accuracy.

The rapidly growing robot market places different demands on sensors than the automotive industry, because higher accuracy, more versatile applications, and output signals that are as standardized as possible are required. However, there is not such a hard cost pressure here. In today's automation technology, positions or rotary movements must be detected above all. But the new generation of robots requires greater sensitivity, that is, not only do the sensors have to be more accurate, they should also be able to "detect with feeling" and therefore require a suitable force sensor in the gripping system. For the detection of workpieces, work tools, surfaces, contours, etc., image sensors are required that can also be used to accurately measure distances. This is more important than ever, especially in modern sensor technology, because the possible applications include electronic, electromechanical, pneumatic, and mechanical measuring tasks.

1.1.2 Basic Metrological Concepts

The measuring means determining the number of units contained in the value of a measured variable:

$$\text{Measured value} = \text{Measurement} \times \text{Unit}$$

The number of units is calculated either from absolute zero or from an agreed reference point of the measured variable. The best-known example of a different counting and naming is the temperature with the units °C (degrees Celsius) and K (Kelvin).

Table 1.2 Basic units of the international SI system

Basic parameter	Base unit	
	Name	Character
Length	Meter	m
Time	Second	s
Weight	Kilogram	kg
Electrical current	Ampere	A
Temperature	Kelvin	K
Quantity of substance	Mol	mol
Light intensity	Candela	cd

For measurement, you need not only a measuring device but also the corresponding units. Today, only the "Système International d'Unités", abbreviated "SI", which was agreed upon internationally in 1960, is used. The SI units are based on seven basic units, as shown in Table 1.2.

The definitions for these basic units are

- Meters: 1 m is the length of the distance that light travels in a vacuum during the time interval of $1/299{,}792{,}458$ s
- Second: 1 s is $9{,}192{,}631{,}770$ times the period duration of the radiation of the nuclide caesium ^{133}Cs.
- Kilogram: 1 kg is the mass of the international kilogram prototype of a platinum-iridium cylinder stored in Paris.
- Ampere: 1 A is the strength of a direct current which flows through two long, straight conductors with a very small circular cross-section running parallel at a distance of 1 m and generates between them the force $0.2 \cdot 10^{-6}$ N/m of their length.
- Kelvin: 1 K is the 273.16th part of the temperature difference between absolute zero and the triple point of water. At the triple point steam, liquid and solid matter are in equilibrium.
- Candela: 1 cd is the luminous intensity with which $1/6 \cdot 10^{-6}$ m^2 of the surface of a black emitter shines perpendicular to its surface at the temperature of the solidifying platinum (2046.2 K) at 1013 bar.
- Mol: 1 mol is the amount of substance in a system of a certain composition, which consists of as many particles as there are atoms in $12 \cdot 10^{-3}$ of the nuclide carbon ^{12}C.

For each measurable physical quantity, an internationally binding unit of measurement and weight is defined within the framework of the international general conferences on weights and measures. These definitions are chosen in such a way that the units are not dependent on material properties or subject to any disturbing influences. In addition, the units form a coherent system in which their relationships to one another are described exclusively by those unit equations which do not contain a numerical factor other than "1".

The most important basic units in electrical engineering are the ampere and the second. Their definitions are:

- 1 A is the strength of a temporally unchanging electric current which, flowing through two straight, infinitely long conductors of negligible circular cross-section, arranged parallel to each other in a vacuum at a distance of 1 m, would electrodynamically generate the force $0.2 \cdot 10^{-6}$ N per meter of conductor length between these conductors.
- 1 s is 9,192,631,770 times the period duration corresponding to the transition between the two hyperfine structure levels of the ground state of atoms of the nuclide Cesium ^{133}Cs.

The main derived units are:

- The unit of electrical power: 1 Watt (1 W) is the electrical power at which energy equal to the mechanical work 1 J = 1 N m is converted in 1 s.
- The unit of electrical voltage: 1 Volt (1 V) is equal to the voltage between two points of a filamentary, homogeneous, and uniformly tempered metallic conductor, in which, with a current of the strength $I = 1$ A, which is invariable in time, the power $P = 1$ W is converted between the two points.
- The unit of electrical resistance: 1 Ohm (1 Ω) is equal to the electrical resistance between two points of a filiform, homogeneous, and evenly tempered electrical conductor, through which, at the electrical voltage 1 V between the two points, a current of the strength $I = 1$ A flows, unchanging in time.
- The unit of frequency: 1 Hertz (1 Hz) is the frequency of a periodic process whose period duration is 1/s.

The exact definitions of the units represent theoretical limits that are only imperfectly and often only indirectly realizable by practical equipment. They are usually not directly suitable for use in a measuring instrument, for example. For example, the exact realization of the unit of current intensity 1 A according to the definition fails because there is no infinitely long conductor with a negligible cross-section.

However, approximate realizations are possible, whose errors can be calculated or whose uncertainties can be estimated. Thus the ampere can be realized with the greatest possible effort with the help of a current balance with an uncertainty of some 10^{-6} Since all other electrical units are now related to the base unit 1 A, it is currently fundamentally impossible to measure any electrical quantity to more than five or six decimal places. However, for some electrical quantities, there are also measuring units which are extraordinarily constant and independent of influences, so that at least changes of a measured quantity or another measuring unit in the order of 10^{-8} or below can be detected.

For most practical measurement tasks, considerably larger measurement uncertainties in the order of 10^{-4} or even 10^{-5} are permissible. For these tasks, easy-to-use and cost-effective measuring standards are available for many storable variables (frequency, voltage, resistance, capacitance, etc.), which can be permanently installed in a measuring instrument as a reference. They are often based on physical principles that differ from the respective definition.

Measuring standards whose structure and properties comply with the certification regulations of the responsible calibration authority (in the Federal Republic of Germany this is the Physikalisch Technische Bundesanstalt PTB) are also called working standards or simply standards.

1.1.3 Analog and Digital Measuring Instruments

In practical measurement technology (measuring instruments at 200 €), a distinction is made between

- Analog measuring instruments,
- Digital measuring instruments.

Analog measuring instruments are pointer instruments and in these, the indication on a scale is made by a pointer. Digital measuring instruments output the measurement result via a multi-digit seven-segment display. The digital measuring instruments are discussed in the second chapter. Figure 1.1 shows the difference between analog and digital measuring instruments.

The analog measuring instrument displays measured values between 0 and 300 V. The digital measuring instrument is a 3 1/2-digit display and shows a measured value of +1.353. While an analog measuring instrument hardly requires any electronics, a digital measuring instrument requires complex additional electronics.

In the case of electrical quantities, an effect is always measured because electricity cannot be perceived directly with our sensory organs, such as the length when measuring a workpiece. The effects of electricity are many and varied and, accordingly, so are the electrical measuring methods. The interaction between electricity and magnetism is most

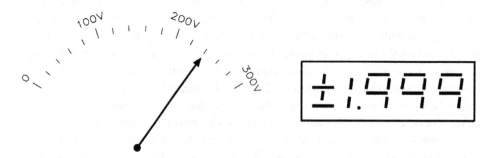

Fig. 1.1 Difference between analog and digital measuring instruments

frequently evaluated. Over 90% of all measuring instruments used in practice are based on the magnetic effect.

In practice, electrical energy can be converted into any other form of energy and its effect can be used to perform measurements:

- Magnetic effect: Every current flow causes a magnetic field and therefore this method is used in 90% of the electrical measurement technology.
- Mechanical effect: In the electrostatic principle, electrically charged bodies of the same name repel each other. The piezo crystal bends when a voltage is applied.
- Heat effect: With the direct effect, the current heats a hot wire and thus changes the linear expansion. If the indirect effect is used, the heated wire is measured by a thermocouple.
- Light effect: A distinction is made between gas discharge and incandescent lamps. The type and length of the glowing light depend on the voltage and the brightness of the filament depends on the electrical power.
- Chemical effect: The amount of gas evolution depends on electrical work.

All measuring devices of this type are based on the physical fact that an electric current causes a magnetic field which depends on the current intensity. If the current to be measured is sent through a coil, a piece of soft iron is drawn into the coil to a greater or lesser extent, depending on the current intensity (Fig. 1.2a).

If the current-carrying coil is rotatably mounted between the poles of a permanent magnet, it rotates against a tension spring, depending on the current intensity (Fig. 1.2b). The dependence of two currents can be measured when the moving coil moves in the field of an electromagnet (Fig. 1.2c). Voltage measurements are also usually based on such current measurements.

Pure voltage measurement is possible with electrostatic methods in which two identically charged plates repel each other (Fig. 1.3a). In contrast to magnetic methods, no current flows, so the measurement is performed without power. Similarly, a mechanical effect can be caused directly by an electrical voltage if the measuring voltage is applied to a special crystal plate, a piezo crystal, which then bends mechanically under the influence of the voltage (Fig. 1.3b).

The flow of current in an electrical conductor generates heat, which in turn can be used as a measure of the strength of the current. Either the linear expansion of wire is measured when it heats up due to the current flowing through it (Fig. 1.4a), or the deflection of a bimetallic strip is measured. Furthermore, the heating can be determined by a thermocouple (Fig. 1.4b). The measuring current is conducted through a resistance wire. A thermocouple touches the wire or sits very close to it. The thermoelectric voltage is a measure of the temperature and thus of the current.

Light effect (Fig. 1.5) is used in some measuring methods by determining the length of a glow discharge or by measuring the brightness of an incandescent lamp, both as a measure of the voltage applied or the current flowing through it.

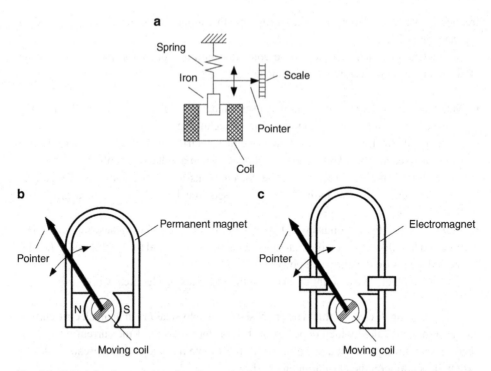

Fig. 1.2 Principle of magnetic action. In moving-iron movements, (**a**) the soft iron piece is drawn into a current-carrying coil. (**b**) In moving-coil movements, the current-carrying coil rotates in the field of a permanent magnet. (**c**) In the electrodynamic movement, the current-carrying coil rotates in the field of a permanent magnet

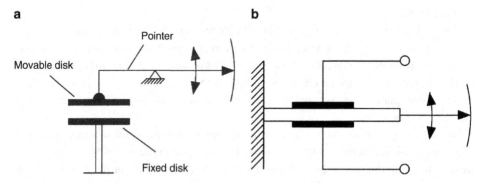

Fig. 1.3 Principle of mechanical action. (**a**) Electrostatic movements repel electrically charged bodies of the same name. (**b**) A piezo crystal deforms when voltage is applied

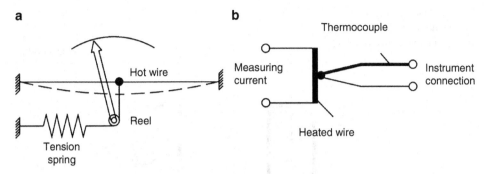

Fig. 1.4 Principle of the heating effect. (**a**) In the hot-wire movement, the current heats the hot-wire and the linear expansion causes the pointer to deflect. (**b**) In the bimetal movement, the current heats the wire and this is measured by means of a thermocouple

Fig. 1.5 Principle of the light effect. (**a**) With gas discharge, the type and length of the glow light is voltage-dependent. (**b**) The brightness of the filament depends on the electrical power

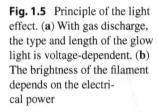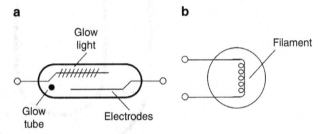

Chemical effects are used by measuring the precipitation of gases or the deposition of metals or salts during electrolysis (Fig. 1.6).

In some cases, the measurement procedure may seem complicated and intricate, but in practice, it is often the simplest principle. It is comparable with energy conversion. For example, the chemical energy of coal is first used to evaporate water, the steam turbine is operated in this way, then an electric current is generated in a generator with the aid of magnetic fields. Nevertheless, this is the more economical process compared to the direct conversion of chemical energy into an electric current in a torch battery. Measurement technology behaves in a similar way. The apparently simplest method of directly converting electrical energy into mechanical motion in the piezo crystal is used only very rarely, whereas the detour via magnetic methods is the most common. Which method is most suitable can only be decided on a case-by-case basis. High demands on the accuracy or low available energy may require special, extraordinary measuring methods.

A measured value must be identifiable, either displayed on a scale or recorded on a recording strip or directly readable in figures. Pointer instruments still account for the largest share of all electrical measuring instruments, although electronic measuring instruments have numerous advantages.

In the course of time, different forms have developed, which have been adapted to the most diverse needs. These shapes were partly due to the physical process, partly to the

Fig. 1.6 During electrolysis, the amount of gas generated depends on the electrical work

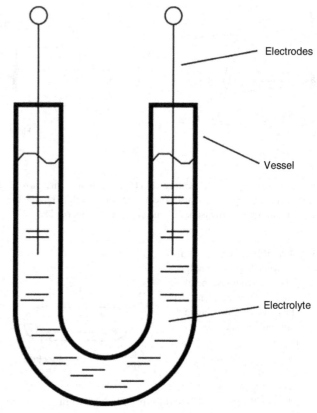

Electrodes

Vessel

Electrolyte

location of the measuring instruments, whether permanently mounted in a control panel or designed as a portable tabletop unit. They were also partly due to the price of the device since an improvement of the display often means a considerable increase in price.

In the case of measuring instruments with a mechanical pointer, the circular arc pointer predominates. This is primarily due to the design of the encoder, since this is the simplest design for the frequently used moving-coil instruments, for example, the moving coil is directly connected to the mechanical pointer to form a unit. If in special cases, a straight and even scale is required, this requirement can be met by deflection or rope guidance. If the front face is to take up as little space as possible, the pointer can rotate in a cylinder cut-out and the unit can be positioned flat behind the control panel.

A certain amount of energy is required to move a mechanical pointer, which is not available in all metrological cases.

An almost inertia-free display is obtained with an oscilloscope or electronic measuring instruments. In the electron beam tube, the beam is deflected magnetically or electrically. Mechanically moving parts do not exist at all. Here it is possible to realize very fast movements carried out and use the measuring instrument as a recorder for very fast processes or vibrations. For simple measurements, the method is too expensive, but for laboratory purposes, it is generally used today.

Parts and accessories of electrical pointer-type measuring instruments (according to VDE 0410):

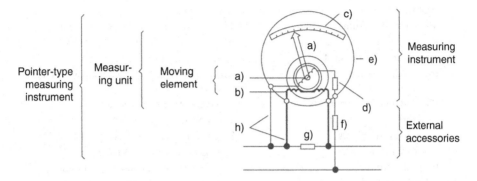

a) Moving element with pointer (e.g. with moving coil in the voltage path)

b) Fixed coil (in current path)

c) Scale = a + b + c = Measuring unit

d) Built-in accessories; e.g. series resistor in the voltage path

e) Housing = a + b + c + d + e = Measuring instrument

f) Separate series resistor

g) Separate shunt

h) Measuring lines = f + g + h = External accessories

Measuring instrument + External accessories – a ... h – Pointer-type measuring instrument

Fig. 1.7 Parts and accessories of electrical measuring instruments

For recorders, the mechanical recording is the simplest. The energy consumption (own consumption) is even higher compared to the mechanical pointer device. The stylus shall be able to overcome the frictional resistance on the expiring paper strip. Low energy consumption and the ability to record fast processes are characteristic of the photographic recording light pen.

To avoid confusion and errors, only standardized names should be used. The standards distinguish between the three important terms measuring mechanism, measuring instrument, and measuring device. Only the moving element with the pointer, the scale, and other parts that are decisive for the function, such as a fixed coil or the permanent magnet, belong to the measuring mechanism. The measuring movement is completed to a measuring instrument by built-in series resistors, change-over switches, rectifiers, and the case. The measuring mechanism alone is functional but not directly usable, the measuring instrument can be used in this form, for example, for tabletop devices. If external accessories are added, such as test leads or separate series and shunt resistors, separate rectifiers, and others, then a complete measuring instrument is assembled. Figure 1.7 shows parts and accessories of electrical measuring instruments. Table 1.3 lists the names of the encoders.

Table 1.3 Designation of the measuring instruments

(a) According to the type of measuring mechanism	Identification
1. Moving-coil instruments	Fixed permanent magnet, moving coil(s)
2. Rotary magnetic instruments	Moving permanent magnet(s), fixed coil(s)
3. Moving iron instruments	Moving iron part(s), fixed coil(s)
4. Iron needle instruments	Movable iron part(s), fixed permanent magnet; fixed coil
5. Electrodynamic instruments	Fixed current coil(s), movable measuring coil(s)
6. Electrostatic instruments	Fixed platen(s), moving platen(s)
7. Induction instruments	Fixed current coil(s), moving conductor(s)
8. Hot-wire instruments	Wire heated by the passage of current
9. Bimetal instruments	Wire heated by the passage of current
10. Vibrating instruments	Vibrating mobile organs
(b) By type of transmitter	Identification
1. Thermal converter measuring instruments	Thermocouple supplies measuring voltage
2. Rectifier measuring devices	The rectifier converts alternating current into direct current
(c) By type of special measures	Identification
1. Quotient meter	The ratio of electrical quantities is measured
2. Summation or difference meter	With two windings, currents are summed
3. Astatic instruments	Paired meter movements with opposing fields
4. Iron-shielded instruments	Iron shielding against external fields

The designation of the types of measuring instruments is also specified in the standards and should be used in accordance with the description. First of all, they are distinguished according to the physical process of measurement (Table 1.3). The measuring instruments are then divided into ten groups. The order and classification is not a value judgment and does not give any information about the suitability of use. It merely states something about the basic characteristics and thus about the possible uses. For example, a moving-coil instrument with a fixed permanent magnet and a moving coil may only be suitable for measuring direct currents. The same applies to the reversal, the moving-coil instrument, in which the coil through which the current flows is fixed and a permanent magnet is movably arranged. Moving iron instruments were often referred to as soft iron instruments in the past. This is obsolete today, as the moving iron part is always mounted so that it can rotate. The iron needle instruments differ from the moving iron instruments by the additional permanent magnet, whose effect is strengthened or weakened by the current flow in the coil. Electrodynamic instruments have a fixed and a moving coil and can therefore display the product of two currents. Instruments with several coils in the moving organ or fixed part also have the same designation. Electrostatic instruments consist of fixed and movable plates. Induction instruments work with currents induced in moving conductors or metal plates. Hot-wire instruments measure the linear expansion of a wire heated by the current flow and bimetallic instruments measure the movement of the heated bimetallic element. Finally, vibrating instruments have mechanically vibrating parts, reeds, or plates that can resonate.

A further subdivision is made according to the type of additional devices for measured value conversion. Conversion is often necessary when alternating currents are to be measured with measuring instruments which are only suitable for direct currents due to their characteristics. Finally, one can also subdivide according to special measures. By attaching several coils in the moving element or fixed part, the formation of quotient values is possible. It is also possible to obtain a sum or difference of two measured values.

In measuring procedures of the magnetic group, external magnetic fields have a strong distorting influence. As a countermeasure, two measuring mechanisms in the astatic instrument can be coupled in pairs so that the external influences cancel each other out. An external field influence can also be eliminated by magnetic shielding.

For quick orientation about the data and properties of an existing measuring instrument, abbreviations and symbols are entered on the scales. These symbols must not be used as circuit diagrams in circuits and circuit diagrams. The symbols are usually grouped together on the scale and must be familiar and commonplace when working with measuring instruments.

The first group indicates the type of current for which the meter can be used (Fig. 1.8). A distinction is made between direct current (DC), alternating current (AC), and direct and alternating current (AC/DC). In the case of three-phase current, bold type indicates whether one, two, or three movements are installed in the meter, which then operates on a single pointer with a scale.

The test voltage indicates how the construction, the terminal distance, and the insulation are tested. Usually, the test voltage is 2 kV, for simpler measuring instruments, especially in communications engineering 500 V. In this case the test voltage star does not contain any numerical information.

The prescribed operating position must be observed without fail, otherwise, the display accuracy will suffer. Usually, it is only indicated whether the indicator is suitable for vertical installation (in a control panel) or horizontal use, for table-top units. In special cases, for precision instruments, a restriction on the permissible deviation can also be given.

The accuracy class consists of a numerical value between 0.1 and 5. As a rule, it is referred to as the full-scale value. Table 1.4 shows the encoder classes.

The largest group of symbols gives data on the operation of measuring instruments and accessories. The symbols are easy to remember because they simplify the structure. The main groups are further subdivided than in the table of designations. There are separate symbols for simple moving-coil movements with one moving coil and moving-coil movements with crossed coils for measuring ratios (quotients).

The information on accessories includes the measuring transducers and the separate series and shunt resistors belonging to the measuring instrument. Electrostatic or magnetic shielding is specified so that the application can be correctly assessed. In some cases, a

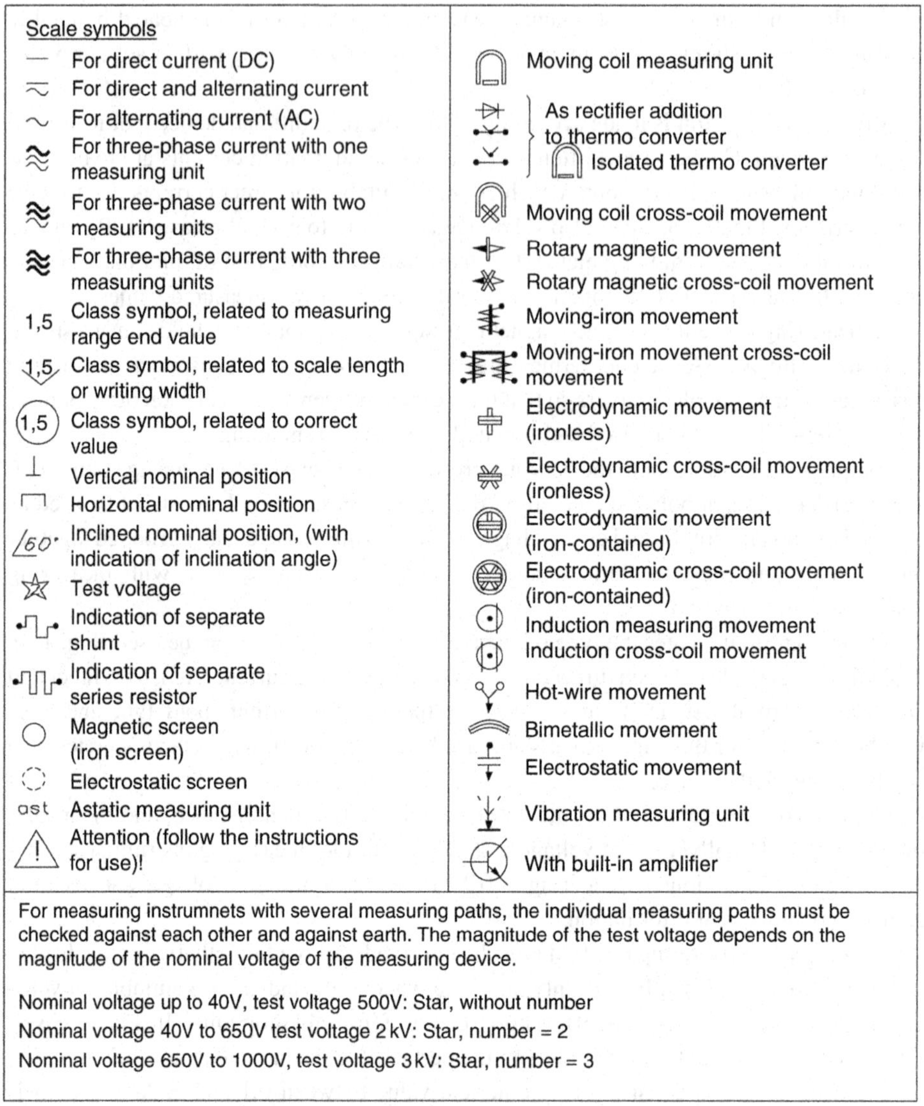

Scale symbols

—	For direct current (DC)
≂	For direct and alternating current
∼	For alternating current (AC)
≈	For three-phase current with one measuring unit
≋	For three-phase current with two measuring units
≋	For three-phase current with three measuring units
1,5	Class symbol, related to measuring range end value
1,5	Class symbol, related to scale length or writing width
(1,5)	Class symbol, related to correct value
⊥	Vertical nominal position
⊓	Horizontal nominal position
/60°	Inclined nominal position, (with indication of inclination angle)
☆	Test voltage
⊓	Indication of separate shunt
⊓⊓	Indication of separate series resistor
○	Magnetic screen (iron screen)
○	Electrostatic screen
ɑst	Astatic measuring unit
⚠	Attention (follow the instructions for use)!

Moving coil measuring unit

As rectifier addition to thermo converter
Isolated thermo converter

Moving coil cross-coil movement

Rotary magnetic movement

Rotary magnetic cross-coil movement

Moving-iron movement

Moving-iron movement cross-coil movement

Electrodynamic movement (ironless)

Electrodynamic cross-coil movement (ironless)

Electrodynamic movement (iron-contained)

Electrodynamic cross-coil movement (iron-contained)

Induction measuring movement

Induction cross-coil movement

Hot-wire movement

Bimetallic movement

Electrostatic movement

Vibration measuring unit

With built-in amplifier

For measuring instrumnets with several measuring paths, the individual measuring paths must be checked against each other and against earth. The magnitude of the test voltage depends on the magnitude of the nominal voltage of the measuring device.

Nominal voltage up to 40V, test voltage 500V: Star, without number
Nominal voltage 40V to 650V test voltage 2 kV: Star, number = 2
Nominal voltage 650V to 1000V, test voltage 3 kV: Star, number = 3

Fig. 1.8 Symbols for electrical measuring instruments

Table 1.4 Measuring instrument classes

	Precision measuring instruments	Operational measuring equipment					
Class	0.1	0.2	0.5	1	1.5	2.5	5
Display error ± %	0.1	0.2	0.5	1	1.5	2.5	5

protective earth connection is provided and specially marked. The zero position for mechanical adjustment of the pointer to the zero marks on the scale is also marked.

In special cases please refer to the instructions for use. In the case of special installation instructions, these are indicated, for example by the instruction to install the measuring instrument in an iron plate of a certain or any thickness. Measuring instruments that are exposed to vibrations have been subjected to a vibration test.

1.1.4 Current and Voltage

Although the base unit 1 A forms the basis for the definitions of all other electrical units, the current is a quantity that cannot be stored and is therefore hardly suitable as a direct measurement standard in a measuring instrument. Unit 1 V, on the other hand, despite its complicated definition, can be permanently embodied by a voltage source and used as a reference when measuring electrical quantities. Current measurements or constant reference currents are therefore also usually referred to as voltage references.

In the 1960s, the alternating current Josephson effect made a process available that makes it possible to generate small direct voltages completely independent of material properties and interference influences and without any drift in time. Figure 1.9 shows a Josephson element consisting of two weakly coupled superconductors that are in almost point-like contact. When the contact surface is irradiated with a microwave of frequency f, it exhibits a staircase DC-DC voltage characteristic with steps of constant voltages in each case, as shown qualitatively. The step heights ΔU are exactly the same and are

$$\Delta U = \frac{h}{2 \cdot q_e} \cdot f \qquad (1.1)$$

The total voltage drop U_J at the nth stage is therefore

$$U_J = n \cdot \frac{h}{2 \cdot q_e} \cdot f \qquad (1.2)$$

In it, h is the Planck's quantum of action and q_e the elementary charge. The proportionality factor between the frequency f and the voltage U_J is an integer multiple of the natural constant $h/2\,q_e = 2.06785\ \mu V/GHz$. With a current of $I_J \approx 4.5$ mA, for example, which does not have to be constant, one reaches the uppermost stage of the characteristic curve. At a microwave frequency of 70 GHz the voltage drop $U_J = 318{,}449$ mV.

Since the frequency f can be measured or controlled with extraordinarily high accuracy and the natural constant $h/2\,q_e$ does not change, it is possible with this method to generate DC voltages in the mV range which are constant to up to ten decimal places and can be reproduced even over long periods of time. By connecting several hundred elements in series, reference voltages in the volt range can be achieved. However, the absolute value of

Fig. 1.9 Operating principle
(a) of the Josephson element
with lö characteristic curve (b)

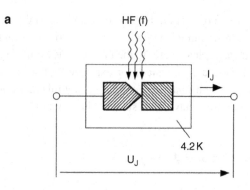

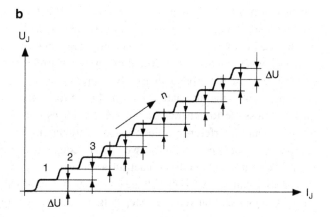

the Josephson voltage is not known more precisely than the constant $h/2q_e$. This is an electrical quantity whose unit refers to the base unit 1 A and can currently be measured to six decimal places at best, as the ampere cannot be realized more accurately.

The Josephson element is not suitable for use in individual measuring devices due to the high expenditure involved. Instead, the international Weston element (Fig. 1.10) has been in use for many decades as the working standard, which chemically generates a DC voltage of 1.01865 V (at 20 °C). It is characterized by a very low and uniform temporal drift, with more than 50% of all working standards ready for dispatch it is below 10^{-6}/Jahr. In the case of thermostatically controlled groups of standard elements, which are used in larger state institutes, such as the PTB, to maintain the unit of voltage, the mean time drift is under $2 \cdot 10^{-7}$/Jahr. It can be measured by means of the Josephson effect.

Disadvantageous properties of the standard elements are their mechanical sensitivity, their relatively large temperature coefficient, which at 20 °C is about $-40 \cdot 10^{-6}$/K, and the fact that they must practically not be loaded.

All these disadvantages can be overcome by the use of reference diodes, which are almost exclusively used in standard measuring instruments today as the measuring standards of the voltage unit. Like all semiconductor components, they have the disadvantage

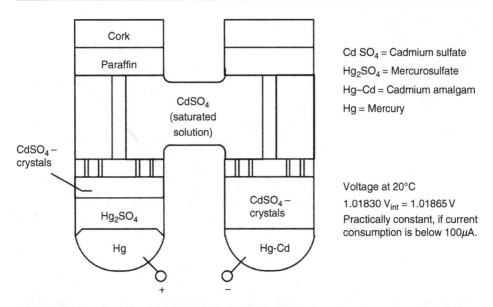

Cd SO_4 = Cadmium sulfate

Hg_2SO_4 = Mercurosulfate

Hg–Cd = Cadmium amalgam

Hg = Mercury

Voltage at 20°C

1.01830 V_{int} = 1.01865 V

Practically constant, if current consumption is below 100μA.

Fig. 1.10 Structure of the international Weston element for a DC voltage of 1.01865 V (at 20 °C)

Fig. 1.11 Time response of reference voltage sources

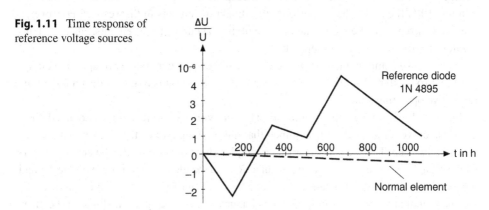

that spontaneous, unpredictable fluctuations of the component parameters impair long-term constancy. Even with the most accurate reference diodes, the long-term stability is at least one order of magnitude worse than that of standard elements. Figure 1.11 shows the irregular fluctuation of the reference voltage of an ultra-stable reference diode, compared to the small uniform drift of a typical utility standard element.

Reference diodes are Z-diodes selected and specified with regard to optimal time response and additionally temperature compensated. Z-diodes can only be manufactured in silicon. Figure 1.12 shows the dependence of the temperature coefficient on the break-down voltage for uncompensated Z-diodes. The temperature coefficient is positive above about $U_Z = 7$ V and can be compensated here by connecting one or more Si diodes in series since their forward voltages each have negative temperature coefficients of about

Fig. 1.12 Temperature
coefficient of Z-diodes as a
function of the breakdown
voltage U_Z and its compensation by 1, 2, or 3 diodes
connected in series

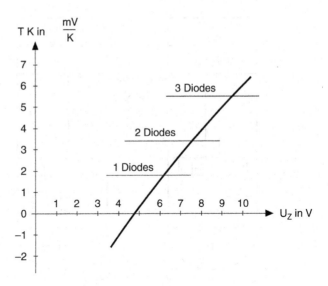

−1.8 mV/K. In this way, the temperature dependence can be reduced by at least one order
of magnitude. In the best case, reference diodes achieve relative temperature dependencies
of a few 10^{-6}/K and are thus superior to the standard elements in this respect. However, a
specified current must be maintained for the highest accuracy requirements, since the temperature coefficient is current-dependent.

As a passive component, a reference diode itself cannot generate a voltage. In a circuit
powered from external sources, it provides a voltage drop that is largely independent of the
respective operating state.

Figure 1.13 shows a simple stabilization circuit with a Z-diode and its associated characteristic curve. First of all, it is assumed that the working part of the characteristic curve
is infinitely steep. Then, as the operating examples show quantitatively, the constant reference voltage of 7 V is at the load resistance R_L, independent of the input voltage U_e and
independent of the load resistance R_L, if one has a circuit for reverse stabilization. It can
also be seen that when the operating state changes, the operating point shifts on the characteristic curve. This is only permissible within certain limits. If the input voltage and/or
the load resistance become too small, the operating point will run into the lower curve
bend and stabilization is no longer possible. The upper limit is given by the maximum
permissible power loss of the reference diode.

In fact, the working range of the characteristic curve has a finite gradient. The reciprocal value, the differential resistance

$$r_Z = \frac{dU}{dI} \tag{1.3}$$

is the output resistance of the stabilization circuit and a measure of the effect of the reverse
stabilization. Values up to about 1 Ω are reached at breakdown voltages of about 5 V.

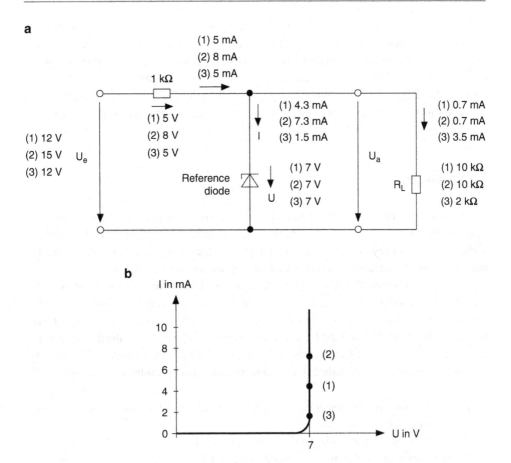

Fig. 1.13 Stabilisation circuit with a reference diode and an operating example and increased input voltage compared to reduced load resistance R_L (1). The characteristic curve of the reference diode with the position of the operating points for the examples

In the example shown, a differential internal resistance of 1 Ω causes the output voltage to increase by 1 $\Omega \cdot (7.3-4.3$ mA$) = 3$ mV when the input voltage of the circuit increases from 12 to 15 V. A measure for forward stabilization is the stabilization factor σ. It is the ratio of the relative changes in the input voltage U_e and the output voltage U_a

$$\sigma = \frac{\dfrac{\Delta U_e}{U_e}}{\dfrac{\Delta U_a}{U_a}} \tag{1.4}$$

The stabilization factor in the example is

$$\sigma = \frac{3V}{12V} : \frac{3mV}{7V} = 583$$

In the example, the diode current is dependent on the operating state of the circuit. Later it will be shown how it is possible to keep the reference diode at a prescribed operating point regardless of the operating state. With such circuits, it is also possible to achieve considerably smaller output resistances up to the order of magnitude of some $\mu\Omega$ and considerably larger stabilization factors up to some 10^5. Finally, it is also possible by electronic means to adjust the output voltage to be stabilized independently of the value of the reference voltage.

1.1.5   Resistors

Resistors are suitable for the direct and alternating current almost without restrictions. As passive components, resistors are particularly easy to store and are therefore ideal as measuring standards. They are required, for example, for measuring active and reactive resistances and, in conjunction with reference voltages, for measuring current.

The most precise possibility to realize a calculable resistance independent of any disturbing influences is offered by the quantum hall effect (v. Klitzing effect) discovered in 1980. This represents a special feature compared to the classical Hall effect. According to the equation, the ratio of the Hall voltage U_H to the control current i_S, which has the dimension of a resistor and is also called Hall resistance R_H, is proportional to the magnetic flux density B. This proportionality is disturbed if the following three special features are fulfilled:

1. The charge carriers representing the control current move in one plane as a two-dimensional electron gas.
2. the magnetic flux density B is greater than approx. 2 T.
3. the operating temperature is near absolute zero at 1 K.

As Fig. 1.14 shows, the $R_H(B)$ curve then has steps whose levels are R_{H_i} integer fractions of the constant $h/q_e^2 \approx 25{,}813\ \Omega$

$$R_{H_i} = \frac{1}{i} \cdot \frac{h}{q_e^2} \quad \left(\text{with } i = 1,2,\ldots\right) \tag{1.5}$$

These resistance values are R_{H_i} absolutely constant, since they depend only on natural constants, and are therefore suitable for determining even minimal changes in other measuring resistances by comparison.

The practical embodiment of the unit of measurement 1 Ω and of decimal multiples or parts of the unit is provided by normal resistors or groups of normal resistors, which are usually made of manganin and have very low time and temperature dependencies. The time drifts are between some 10^{-5}/Jahr and less than 10^{-7}/Jahr, depending on the effort involved. The temperature coefficient is some 10^{-6}/K at the given nominal temperature (mostly 20 °C). Normal resistances are also designed to contain the lowest possible reactive resistance components and to have the highest possible load capacity. As shown in

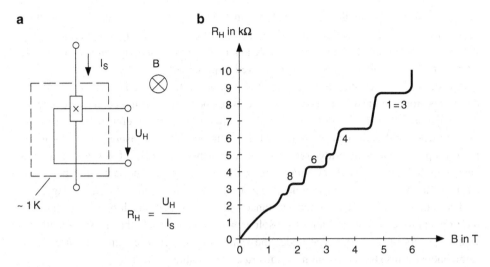

Fig. 1.14 Quantum Hall effect. (**a**) Hall sample. (**b**) Hall resistance R_H as a function of magnetic flux density B

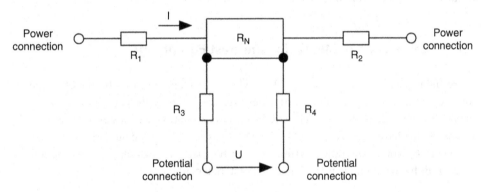

Fig. 1.15 Connections of a normal resistor with RN as normal resistance, $_{R1,...,R4}$ as undefined line and contact resistances

Fig. 1.15, they are always provided with separate current and potential connections to avoid measurement errors due to parasitic voltage drops at the supply and contact resistances R_1 to R_4.

For the embodiment of the units of measurement of alternating current resistances, normal capacitors, normal inductances, and normal counter-inductances are also available, which do not play a significant role in measurement electronics and which will therefore not be discussed in detail.

Standard resistors are independent device-like units and are not suitable for permanent installation as reference resistors in measuring instruments. For this purpose very precise metal film resistors are available, which also reach temperature coefficients of a few $10^{-6}/\mathrm{K}$. However, metal film resistors are less stable over time than standard resistors. For the highest demands, smaller resistor values can be manufactured from manganin itself up to about 10 Ω. This material has the advantage that it has particularly low thermoelectric voltages of the order of magnitude 1 µV/K against copper. This plays an important role wherever low DC voltages have to be measured or processed.

The carbon film resistors most commonly used in electronics and especially semiconductor resistors are unsuitable as reference resistors because their temperature dependencies in the order of 0.1%/K and their time drifts are too large for reference purposes.

In practice, only constant resistance ratios are important in measurement technology. For example, a measuring bridge remains balanced if the resistances of its bridge branches change in the same ratios. Precise resistance ratios also play an important role when comparing different voltages, for example, a measurement voltage with a normal voltage.

The requirement for a constant resistance ratio is considerably less critical than the requirement for constant resistance to absolute values. It can usually be met sufficiently by using resistors of the same type under the same environmental conditions. Often even semiconductor resistors on a common substrate are suitable.

Extremely precise resistance ratios with uncertainties of a few 10^{-9} can be achieved with inductive voltage dividers, whose divider ratios are only determined by the number of turns and are therefore independent of material properties.

1.2   Calibration of Measuring and Test Equipment

More and more often the question is asked, which advantages and disadvantages the often much cheaper factory calibration certificates offer compared to the calibration certificates of the German Calibration Service (DKD). Works calibration certificates are issued with standards that have been calibrated by DKD offices or verification authorities. As shown in Fig. 1.16, works calibration certificates can be issued by external calibration bodies (such as B. by the manufacturer), but also by internal calibration laboratories.

Fig. 1.16 Structure of the calibration services

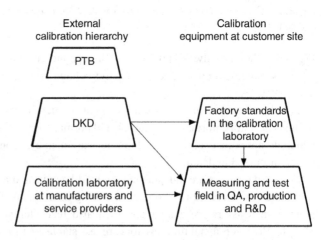

The calibration certificates of one of the calibration laboratories represented in the German Calibration Service (DKD) and NIST Organication guarantee the highest level of measurement reliability. Standards of these calibration laboratories, such as of DKD 2101, are recalibrated directly at the Physikalisch-Technische Bundesanstalt (PTB). The PTB accredits the DKD bodies and only regular auditing by the PTB entitles the holder to bear the DKD seal. High-quality measuring equipment, the precisely defined climatic environmental conditions, and the strict calibration requirements of the PTB stand for the quality and reliability of the calibration certificates issued here. These include B. the prescribed repeat measurement of each measuring point at longer intervals to increase safety, which is achieved by averaging two measured values.

The creation of works calibration certificates on-site by means of own works standards is certainly useful, and even for measuring and test equipment that is not critical in terms of calibration accuracy, the works calibration certificate can often be a cost-saving solution.

On the other hand, the recalibration of factory standards, with which another measuring and test equipment is to be calibrated, can only be advised to obtain a DKD calibration certificate due to the high measurement reliability required. And also measuring and testing equipment, on whose results much depends, should be recalibrated with a DKD calibration certificate if in-house factory standards are not available.

1.3 Analog and Digital Data Acquisition

A measured value can be communicated to the user in various ways: on a scale, in digits on a display, as a waveform in an oscilloscope, or by outputting it via the monitor or printer of a PC system. Until about the mid-1980s, pointer-type measuring instruments accounted for the largest share, which were then partly replaced by the less expensive digital measuring instruments. While a pointer instrument requires a higher degree of technical understanding—because you have to be able to read a scale correctly—digital measuring instruments work largely "foolproof". When digital measuring values are recorded with a PC system, not only can one or more measured values be displayed on the monitor or printer, but they can also be stored and, if necessary, numerically processed, in order to output them graphically, e.g. according to individual design.

1.3.1 Structure of an Analog Electrode

In the analog measuring instruments with a mechanical pointer, the circular arc pointer prevails. This is primarily due to the design; the simplest design is the much-used moving-coil instrument. The mechanical pointer is attached directly to the moving coil. If a straight scale is required in special cases, the rotary movement is deflected by means of a cable guide. The movement of a mechanical pointer requires a certain power; if this is not

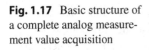

Fig. 1.17 Basic structure of
a complete analog measurement value acquisition

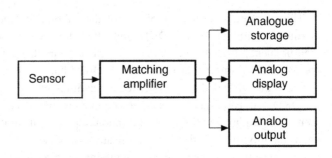

available, an amplifier is required. Figure 1.17 shows the principle structure of a complete analog measurement acquisition.

At the beginning of the analog measuring, the chain is the sensor, which converts the physical, chemical, or geometrical quantity into an electrical quantity. This is followed by the matching amplifier (which usually works with operational amplifiers) and is terminated by the analog display, the analog output, or the analog storage. The storage of analog signals is difficult. In the past, expensive magnetic tape recorders were used, but this recording technique is obsolete today.

With an analog output, the input signal is amplified to 1 V output voltage. These amplifiers usually also act as impedance converters with high input and low output resistance. With MOSFET operational amplifiers, which can have input impedance up to 10^{25} Ω while their output impedance (standard impedance is usually 60 Ω) is very low, even very high-impedance sensors can be adapted to almost any subsequent circuit.

A very frequently used sensor interface is the current interface with a constant current from 4 to 20 mA. A current value from 4 mA corresponds here to a. A measured value of zero; if the evaluating circuit detects a lower current, it interprets this as a defect (e.g. Line interruption). A higher current than 20 mA indicates an over-range or a short circuit.

1.3.2   Structure of a Digital Measurement Chain

The simplest case of digital evaluation is when the sensor signal is directed to a comparator which outputs the value "1" when the threshold value is exceeded and "0" when it falls below. To prevent the output from switching too often, in practice the switch-on and switch-off points are set to slightly different values (which can be set with a potentiometer, e.g.); the distance between the two is called hysteresis. Such a circuit is a "1-bit" AD converter, so to speak.

The principal scheme of a digital signal processing is shown in Fig. 1.18. The sensor generates a measured value which is generally brought to a voltage level in the Volt range by an adaptation amplifier optimized for the respective application. This is then converted into a digital value by an analog-to-digital converter. From here on, a distinction must be made between digital measurement technology and PC measurement technology. If the

Fig. 1.18 Principle structure of a complete digital measured value acquisition

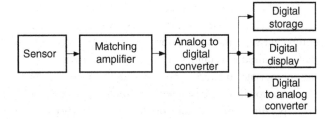

measured value is output directly via a seven-segment display, you have a digital measuring instrument. Digital storage is only possible in conjunction with a PC system. Since the beginning of the 90s there are also digital multimeters that can be connected to a PC via a serial interface. In addition to the digital display, one now also has a monitor or printer available for registration.

There are many known methods for digitizing an analog input voltage. Since a special A/D module has been developed for each conversion method, each individual A/D method can be used advantageously under certain application conditions. In addition to the fundamental errors that arise during conversion, each conversion method also involves system-related errors. Therefore, a basic knowledge of the different conversion methods is advantageous for the user. For a better overview, the different procedures are divided into three groups:

- The first group includes all converters with an indirect mode of operation, i.e. during the conversion, they first generate an intermediate signal from the analog input signal, which is converted into the final result in a second step.
- The second group includes all A/D converters that digitize the input signal directly but require one decision element (comparator) per quantization step (resolution).
- All directly converting AD converters are assigned to the third group but only need one decision element.

1.3.3   Acquisition and Processing of Measurement Data

Nowadays, the acquisition and further processing of measurement data are hardly common without a connection to a PC system. This has led to the fact that sub-processes, which in the past were handled by different departments with different tools, are now combined with one and the same medium, possibly using one and the same software, into a single, integrated process. The planning of measurement tasks, the realization of sequence controls, the monitoring of processes, the control of test benches, or the statistical analysis of raw data are thus no longer separate tasks but are solved holistically. This increases the transparency of the system, improves communication between departments, shrinks hierarchical levels, and saves enormous costs.

The aim is therefore to achieve integrated solutions; they make previously required special acquisition devices, recorders, evaluation devices, and subsequent documentation equipment superfluous and almost always far exceed them in terms of performance. All that is required is a standard PC, acquisition hardware (AD card or front-end system), and the appropriate software for processing the measurement data.

The following use cases show examples of how flexibly extended PC hardware and the corresponding software can work together: Slowly changing process data such as temperature, gas quantities, light, torsion (mechanical twist), or pressure changes were often stored and recorded by continuous recorders in the past. In many cases, a more economical replacement can be found by means of software. A program part of the measurement processing software can be used to simulate an endless recorder "online", whereby any normal printer that can process endless paper can be used as an output device. The acquired data are not only output on the printer but are also temporarily stored on the internal hard disk. In this way, measurement data already stored can be output again as a copy. Very fast processes can be viewed via a program section in a special "Scope" mode. Just as on a conventional oscilloscope, the individual trigger signals can be set "online" in the "Scope Display", which is directly displayed in the monitor, the trigger threshold can be continuously adjusted, an amplitude comparison can be made or the time base can be changed.

Today's software products for measurement data processing offer a wide variety of solutions for almost all measurement tasks, such as controlling, testing, and monitoring processes, or comparing and evaluating procedures. Signals (measured variables) are recorded via sensors and converted into voltages or currents, amplified and conditioned for the computer with an AD converter. Figure 1.19 shows a solution from the converter to the output on the screen.

The digitized signals are further processed by the measured value processing software. This includes, for example, calibration to obtain the physical quantities again (e.g. temperature), online calculation and display of the measured values, and finally mathematical offline analysis and documentation.

In online operation, the peripheral devices are connected to the computer; this provides the basic requirements for dialog operation and real-time processing. Here the data is entered directly and processed immediately, there is a constant change between input,

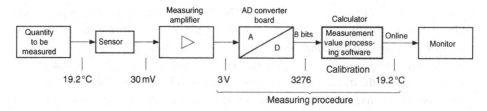

Fig. 1.19 Solutions of different measurement tasks using standard hardware of a PC system, a special card for the acquisition of the measuring instruments with conversion, a software for processing the measured values, and a monitor for the output of the information

processing, and output. Real-time processing guarantees a high level of data and information actuality of the real processes but requires higher expenditures in the area of hardware and software.

The opposite of real-time processing is the offline operation or batch processing. Here, the data is processed with a time delay to the processes taking place in reality. They are first collected (stacked) and only then processed in a program sequence, which can be very fast with a PC system in the upper price range. The user can no longer intervene in the actual process, as is possible with dialog processing. Offline data processing is less expensive than online processing, but cannot offer the same high level of data timeliness.

In addition to the external tasks and the required solutions, the software for measurement value processing must be oriented towards a further area of the environment. This includes both the PC hardware and, in particular, the function in connection with the measurement hardware usually used in measurement technology.

In the field of PC hardware, this means that most data acquisition programs run on PCs with the Windows operating system. This PC hardware can be easily connected via network applications with each other and can therefore also be used for flexible measurement data processing. For example, the programs for measurement data acquisition can be used very mobile in conjunction with a laptop for "on-site" measurements. The measured data is then transferred directly to a central computer via a network. Here, the data can be analyzed while the measurement is still running using a program for data acquisition with special programs. In this case, the computer serves as a control center with decentralized and autonomous measurement data acquisition in the test field.

In the field of measurement hardware, both PC plug-in cards, which have been in heavy use for several years now, and measurement devices with numerous digital interfaces have long been established on the market. The communication of this measurement hardware with the measurement software is done via drivers. Today, the user has a wide range of drivers to choose from. A driver program is a special software for adapting a general problem to user software. If a driver for special measurement hardware is not available, it can be created by the user via a comfortable driver interface.

The PC-supported acquisition, monitoring, control, and regulation of processors run in real-time, i.e. there is a constant flow of data between the acquisition hardware and the measurement software, and the processing of the data with this software is so fast that it can constantly follow the real changes in the process. This places high demands on the software in terms of performance and range of functions, which can only be met with a flexible program structure.

Simplified is the whole program concept for a data acquisition software by using the terms "data", "core" and "driver" as described in Fig. 1.20. In the data area, the measured values are stored in the data matrix or in a file in such a way that this part of the program can access them directly for output or for offline analysis. The principle of saving the measured data in files is designed in such a way that even in the event of a computer failure during writing to a file, the previously measured data will not be lost.

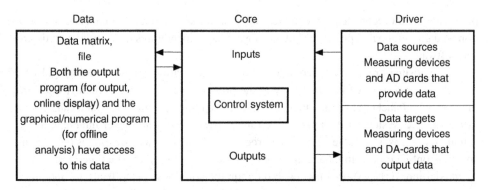

Fig. 1.20 Simplified program concept for a data acquisition software for measurement and control with the individual elements "data", "core" and "driver"

The core describes the hardware-independent part of the measurement task. In addition to the inputs and outputs, it also contains the central control elements that enable the interaction of data and drivers. The hardware-independent drivers provide the connection between data sources or targets and the core. In addition to AD converter cards and front-end devices, these can also be e.g. data channels, the keyboard, or the PC loudspeaker.

The control elements of the core are procedures, conditions, and calibrations. Together with defined conditions, the processes control the temporal course of the incoming and outgoing data streams. They can also be scaled. In the simplest case, the sequence can consist of "start", "execution" and "stop" of the measurement. In special data acquisition programs, however, almost any complex and nested sequences can be defined using conditions, which can be used to start and stop a measurement, to switch sampling rates, to switch inputs and outputs on or off, or to change the online display, for example, a. In addition to conventional conditions such as triggering on windows and edges, these programs also contain rise, time, and sample conditions as well as free mathematical and Boolean conditions. Mathematical functions include all mathematical conditions, while Boolean functions are logical operations.

Linear, non-linear (for thermocouples according to DIN IEC 584), and freely definable procedures are available for calibration. Externally amplified signals can be linearly pre-calibrated. In addition, a calibration measurement can be performed to tare measurement channels.

Figure 1.21 gives an example from practical PC measurement technology. A test bench is to be monitored with regard to sound pressure level and oil pressure. It is always switched off automatically when the condition "true" becomes true that a threshold named "Level 1" is exceeded and at the same time a threshold named "Pressure 1" is undercut or the threshold named "Level 2" is exceeded or the threshold named "Pressure 2" is undercut. In addition, the test bench can be switched off manually by simultaneously pressing the function keys F2 and F3 on the keyboard.

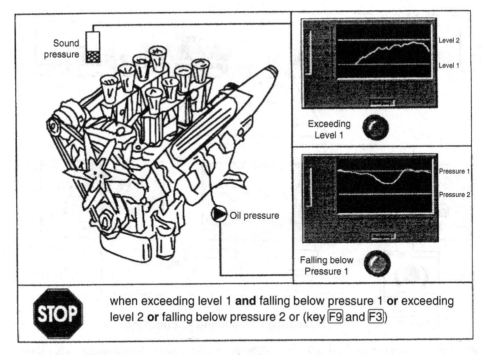

Fig. 1.21 Example of online monitoring of an engine test bench. The test bench is switched off when preset threshold values are reached or keys are pressed

The program core distinguishes between software- and hardware-controlled measurements. The former are subject to complete control by the core so that extremely complex measurement sequences (e.g. Test bench controls) are possible. Software-controlled measurements allow total sampling up to approx. 1 MHz (hardware-dependent!), whereby up to two billion measured values can be recorded. In contrast, hardware-controlled measurements allow data to be acquired very quickly (up to the GHz range). This requires special driver routines that exploit the specific capabilities of the plug-in cards and devices and allow disk, high-speed, and/or DMA measurements.

1.3.4 Control, Regulation, and Visualization

In many areas of process monitoring, there are control processes that are in the low-frequency range and can therefore be solved by software. The PID controller integrated into the program package has proven its worth in solving control and process monitoring tasks, including a. in the areas of temperature, speed and torque monitoring, air conditioning technology, engine test benches, and pressure control.

In general, the aim of a control system is to maintain certain variables at specified setpoints. Disturbances which affect the process should have as little effect as possible on the

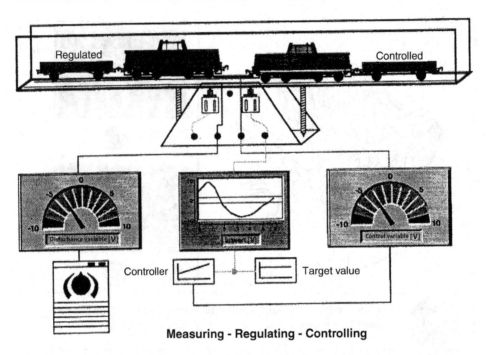

Measuring - Regulating - Controlling

Fig. 1.22 Example of a control system: the right locomotive is moved by the operator via a transformer so that the rocker loses its balance. The software regulates the left locomotive so that the rocker reaches equilibrium again

variables to be controlled. For this purpose, general control algorithms are used in the software packages which determine the manipulated variables (the difference between actual and setpoint) from the measured actual values (any inputs) and the specified setpoints (constants, data channels, measuring channels). These are scaled back accordingly and applied to the outputs.

Auto sequences also enable comprehensive optimization strategies (for example, according to Ziegler-Nichols, Hrones, and Reswick), which facilitate the selection of the optimum control parameters. Figure 1.22 shows an example of a rocker that is to be kept in equilibrium. The right train, the disturbance variable, is moved by the user via a transformer, while the left train is moved by the PID controller of the PC system so that the rocker is rebalanced.

Especially important for all groups of users are the visualization possibilities of measuring processes on the screen. Anyone preparing a measuring task must find functions here that they can use in the simplest way to present their measuring process in the best possible way. For the potential end-user, the visualizations must be clear and concise and correspond to the usual presentation methods. The spectrum of visualizations ranges from time-domain displays, numerical displays, pointer displays, polar coordinates to binary display forms. A special feature of the online graphics is the integration of static background graphics.

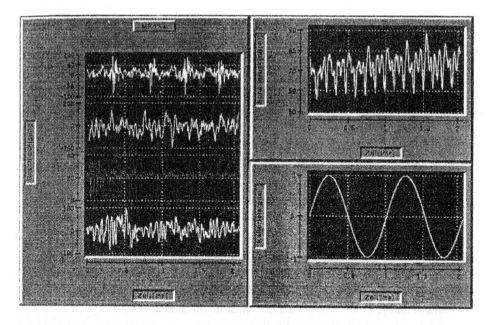

Fig. 1.23 PC system as an oscilloscope. The representation depends on the number of analog input channels of the PC plug-in board and on the possible sampling rates

With the ability to take measurements, change the online display based on conditions, and plot warning and exceedance limits, the software allows complete processor simulations. It is also possible to write to the RAM or directly to the hard disk in parallel to the online display with high acquisition rates in the form of a file. The actual acquisition of data has higher priorities than the online display so that all data are acquired in any case, even if not all of them can be displayed immediately.

Rapidly changing signals can be viewed on the monitor of a computer. This requires not only the hardware but also software with a special program part, as shown in Fig. 1.23. Just as on a conventional oscilloscope, the Scope mode allows the various trigger conditions to be set online, the trigger thresholds to be continuously changed, an amplitude comparison to be made and the time base to be changed. The maximum possible sampling rate is only limited by the hardware used.

All incoming measurement data can already be evaluated during the running measurement with different algorithms. The simplest form is the direct calculation (e.g. Basic calculation functions, trigonometric functions, Boolean operations) of different measuring channels. During the measurement, the software then links the inputs, constants, or already existing data specified in the formula to new data according to the calculation rule and, if necessary, displays these in the online graphic.

All information for the definition of a measurement setup is summarized in a setup file which is saved under a freely definable name in ASCII format and can then be loaded again. Thus even complex measurement sequence definitions are available again literally

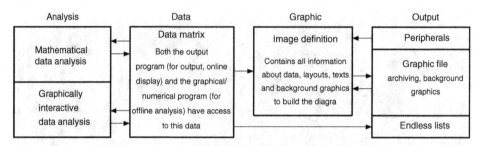

Fig. 1.24 Simplified program concept for analyzing, documenting, and archiving with the elements "Data", "Analysis", "Graphics", and "Output"

at the push of a button for further measurements or for modification for similar measuring tasks.

The advantage of being able to quickly convert, store and analyze large amounts of data by means of digital data acquisition and processing has now led to the problem that users are faced with an almost unmanageable flood of measurement data. With the help of a PC and standard software programs, this can be made clear once again, far exceeding the conventional analysis and documentation of measurement data.

Figure 1.24 shows a highly simplified program concept for analyzing, documenting, and archiving with the elements "Data", "Analysis", "Graphics" and "Output". The data, which are managed in the software in the form of a data matrix, can come from direct PC measurements, from external files, or from manual entries. As is usual in measurement technology, the program for measurement data acquisition starts channel-oriented, whereby more than 1000 channels allow more than two billion values to be processed simultaneously, if the corresponding hardware and software is available.

The data form the basis for both the mathematical offline analysis and the graphics. In the area of analysis, numerous mathematical functions and procedures are available as well as the graphically interactive analysis. The central element of the graphic is the image definition, which contains all information about the structure of the diagram. The output can then be made in the highest quality directly to a printer or plotter, or for archiving purposes to a file in the appropriate graphic format. It is also possible to make endless recordings or to output the data as an alphanumeric list.

An essential and very extensive part of the program is the mathematical generation of data channels and the analysis of raw data. In addition to general calculation functions—e.g. Formula Interpreter, Integration, RMS value—extensive tools for signal analysis are available, e.g. FFT (Fast Fourier Transformation), digital filters, spectrum analysis, coherence, etc. The field of statistics includes both the descriptive functions for determining characteristic values and the inductive functions as aids for quality assurance (e.g. Control Charts) and for statistical process control (SPC). In addition to the usual one-dimensional classification methods according to DIN 45667, there are also more complex methods for classifying data, such as compound classification and the two-dimensional Rainflow method. In the area of curve evaluation, for example, B. with approximation and spline

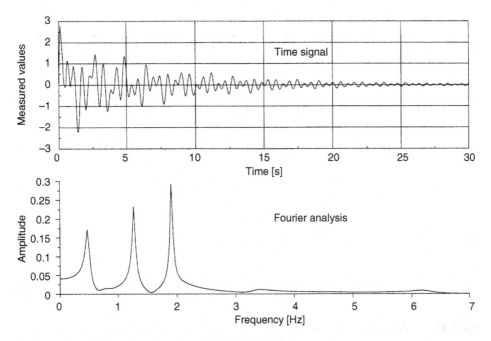

Fig. 1.25 Generation of the time signal and Fourier analysis

functions, the necessary compensation curves can be calculated from raw data. Figure 1.25 shows a time signal with the corresponding Fourier analysis.

In addition to 2D analysis, 3D analysis functions are also available in the program packages. These include multidimensional basic functions for organizing the data structure, e.g. sorting, transposing, conversions, triple-matrix, and operations for matrix linking, as well as 3D evaluation functions such as 3D interpolations and 3D approximations, the calculation of 3D iso-lines (contour lines), and the determination of the integral under a surface.

The skilled user knows that in measurement technology, situations occur time and again in which measurement data are falsified by external events. If the influences are known, corrective action can be taken after completion of a measurement. In the programs, for example, it is possible to work with non-existent values, the so-called "NoValues" in the graphics and analysis part.

If, for example, a sensor was overdriven for a certain time due to external influences, the mathematical analysis will result in a completely distorted histogram. If you have overdriven signals before the histogram calculation, they are automatically assigned the value "NoValue" according to a pre-defined condition. These values are ignored in graphical representations and mathematical analyses so that the static evaluation shows the correct distribution of the measurement data. In addition to mathematical evaluation, these programs also allow a graphical interactive analysis of data in the cursor graphics. Behind this term lies a comprehensive range of functions that allow simultaneous analysis,

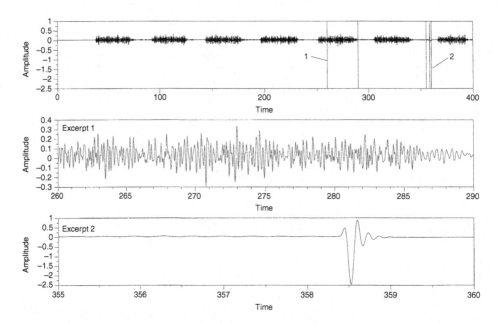

Fig. 1.26 Display of three input channels

especially of long data sets, directly on the screen. Various cursor types (local/global frame/band cursors, crosshairs) are available for convenient use, with which the interesting area of the curve can then be edited.

The various measurement modes are used for the precise measurement of curves. Even editing options for the graphical editing of data sets (e.g. Removing outliers) are available in the program packages. The cursor graphics can also be used, for example, to examine data for certain events. For this purpose, subareas can be zoomed out for more detailed graphical analysis and created as new data sets. These sections can then be subjected to various mathematical analyses, just like the original data.

After the mathematical analysis or the general preparation of the data, they should be interpreted graphically in most cases. For reasons of clarity, the output as a numerical series or as simple curves is now a thing of the past. In contrast, attractive and meaningful documentation of the measurement data is required. Figure 1.26 shows the representation of three input channels.

The software can also be used to quickly and easily create perfect diagrams with any combination of axis systems, tables, and texts. Integrated background graphics illustrate the origin of the data or a measurement setup. Libraries with pre-defined elements (layouts) enable the creation of high-quality documents right from the first start of the program. For the graphical and numerical documentation of XYZ dependencies of the measurement data, the 3D axis systems with and without visibility clarification and 3D tables are available in the programs. With the calculation and display of ISO lines and the color palette display, only two interesting capabilities of 3D documentation are mentioned here.

1.4 Measurement Error

The output signal x_a of a measuring element generally depends on the input signals $x_{e1}, ..., x_{en}$, which are to be linked according to a given instruction, on the internal states $x_{i1}, ..., x_{in}$ (e.g. stored values or switch positions) as well as on the time derivatives of the entire information. Thus, the behavior of the input signal can be quantitatively inferred from the respective value of the output signal, whereby clear and reproducible nominal functions must be defined for these dependencies. These can be in the form of differential equations, diagrams, or tables.

In most cases, these are relatively simple functions. Often an output signal is only dependent on one input signal. Furthermore, relationships are often independent of time and frequency. In these cases, the theoretical behavior of a measuring element can be described by a nominal characteristic curve.

The shape of a nominal characteristic curve depends on the task and the technical possibilities. Often linear or proportional characteristic curves are optimal and it can be the task of electronics to linearize an empirically given non-linear characteristic curve. For some tasks, however, defined non-linear (e.g. quadratic or logarithmic) characteristics are also required.

The theoretically correct (true) value x_w (setpoint) of a measuring signal can be determined for each operating state and time on the basis of the given nominal characteristic curve of a measuring element or its nominal function. Metrologically, the actual value x_a (actual value) of the same signal can be determined under the same conditions. If both values match, the measuring element is error-free in the considered state. In practice, there are no error-free measurements. Even with great effort, an uncertainty remains with every measurement because e.g. a measured value can only be given with a finite number of decimal places. A complete measurement result consists of two pieces of information, i.e. one concerns the value of the result, the other its error or uncertainty.

Measurement error (deviation) A is the difference between the output (or displayed) value x_a of a measurement signal and the true value x_w of the same signal:

$$A = x_a - x_w \tag{1.6}$$

The relative error A_r is the ratio of the measurement error to a reference value X:

$$A_r = \frac{A}{X} \tag{1.7}$$

In this case, the error indication must contain an indication of which is the reference value. The following additions to the error information are common in practice: m.V. (of the measured value), c.V. (from the correct value), R. (reading = from the displayed value), v.E. (from the measuring range end value) or F.S. (full scale = from the measuring range end value).

The errors of a measuring element or measuring device can be of different nature and differ in their origin. Some important influences on the error behavior of electronic measuring equipment, which depend on external parameters or conditions, shall be investigated. The error propagation during signal processing by several measuring elements and when several individual errors occur simultaneously shall also be dealt with.

1.4.1 Types of Errors

When considering the measurement errors, a distinction must be made between three types of error:

- Gross errors are caused by avoidable carelessness during a measurement or during the development of a measuring device and the measurement result is useless in the test. With electronic measuring equipment, gross errors can be caused, for example, by a defective operational amplifier, by a programming error in the software of a microcomputer, or by the incorrect reading of a numerical display. Gross errors must be avoided under all circumstances by suitable checks or multiple measurements.

As an example of random errors, a pointer-type measuring instrument shall be used and shall include

- Fluctuating properties of measuring instruments (loose contacts, cold solder joints, fluctuating contact resistances in the measuring leads) and influencing variables that are difficult or impossible to detect, such as air humidity.
- Reading error due to parallax at the observer. As can be seen in Fig. 1.27, the correct measured value is only read if the eye of the observer is exactly vertical above the pointer. Random reading errors occur when looking sideways. The error becomes smaller the closer the pointer is placed above the scale and the less the direction of view deviates from the vertical.

With precision measuring instruments, a mirror is often placed under the scale to check the vertical viewing direction by aligning the pointer with its mirror image.

Random errors caused by changes in the DUT, its environment (environment, observer), or the instrument itself that cannot be detected or influenced with difficulty have an unpredictable magnitude (magnitude) and sign. Thus, by measuring the same physical quantities several times, different measurement results are obtained due to random errors. By statistically evaluating these results, conclusions can be drawn about the true measured value (nominal value) and the measurement uncertainty.

The most probable value of several deviating measurements can be seen as the average (arithmetic mean) of the individual values. If n independent individual values

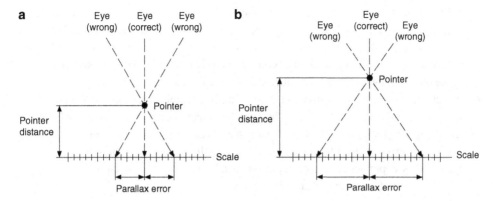

Fig. 1.27 Parallax error when reading the measured value. (**a**) Small distance of the pointer from the scale = small parallax error. (**b**) Large distance of the pointer from the scale = large parallax error

$x_1, x_2, \ldots, x_{n-1}, x_n$ are determined by an observer under the same conditions of a series of measurements, the arithmetic mean value x can be calculated. This probable value is not necessarily the correct value. The closer the individual values of the measurement series are to each other, the more measurement values have been determined, the greater the probability that the arithmetic mean x is the correct value. Note: Individual values are independent of each other if subsequent measurements are not influenced by the preceding one. $\bar{x}$ is an estimated value for the expected value.

• Systematic errors are measurement errors that can be repeated under repeatable operating conditions of the measuring system. They have a certain value and a certain sign in the respective operating condition and are caused by imperfections of the measuring system. The greater the systematic errors of a measuring system, the lower the accuracy. The term "accuracy" is not quantifiable and should only be used for qualitative statements.

Systematic errors can be dealt with in different ways. They can be reduced by appropriate measures, the measurement result can be corrected if the error is known, or error limits can be estimated for the worst case.

• Error reduction: Typical measures to reduce errors in electronic measuring equipment are
  – Linearization of the characteristic curve by electronic means.
  – Adjustment (calibration) of the zero point and the gradient of the characteristic curve. Most measuring instruments have devices for adjusting the zero point, since the measured variable "zero" can usually be easily realized, e.g. by a short circuit. Electronic measuring instruments often have devices for regular automatic adjustment of the zero points. The term "calibration" is often used incorrectly for "adjustment" (see error detection).

- – Elimination of influences acting on the measuring system from outside, or their compensation through oppositely acting dependencies on the same influencing variables.
- – Decoupling of measuring elements and measuring devices from each other and from the object to be measured to reduce the feedback effects.
- Defect recording: Since systematic defects are repeatable, they can be measured in most cases with reasonable effort. They can be detected by a calibration in which the actual dependence of the quantity of interest x_a on an input quantity x_e or an influence quantity is examined with the aid of a sufficiently accurate comparison measuring device. This dependence can be represented as a measured characteristic curve and compared with the nominal characteristic curve.

Often the systematic errors are so small that both characteristic curves practically coincide. In these cases, it is better to represent the systematic error as a function of the associated input or disturbance variable in an error curve. With the aid of the errors thus determined, it is possible to correct the measured value obtained or a signal value by subtracting error A from the output value x_a according to the equation to obtain the true value x_w. This can also be automated with the aid of an electronic computing device.

Figure 1.28 shows in the upper diagram the proportional nominal characteristic of a measuring element and the measured characteristic, which is progressive and does not pass through the zero points. The lower diagram shows the associated error curve, which in this case is also called the linearity error curve. An important parameter is the linearity error of this curve and this is the (in this case: negative) maximum linearity error A_L and the (in this case: positive) zero-point error A_0.

- Error estimation: If it is technically impossible or too costly to determine the systematic error of a measuring instrument or measuring system for each operating case, one is satisfied with the specification of the absolute or relative measurement uncertainty (u_s and ε_s are systematic components), which will not be exceeded even in the worst case. The actual errors A and A_r are then always smaller than these limits:

$$|A| \le u_s \tag{1.8}$$

$$|A_r| \le \varepsilon_s \tag{1.9}$$

For the maximum errors of electronic measuring instruments, the manufacturer specifies error limits which constitute a guarantee. For example, the manual of a digital voltmeter states

$$\text{Error smaller than} \pm 0.03\% \text{v.E.} \pm 0.06\% \text{m.v.} \left(\text{from measured value} \right)$$

Fig. 1.28 Nominal characteristic, measured characteristic
and error curve of a measuring
element with A_0: Zero point
error A_L: Maximum linearity error

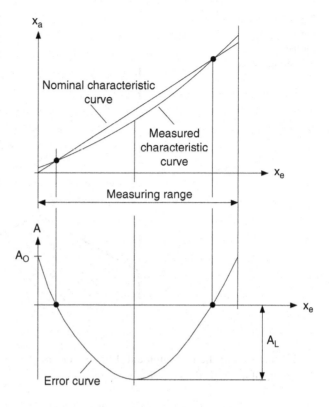

This means that with a measuring range from -2 V to $+2$ V the following values are obtained

$$\text{Zero point error}: |A| \leq u_{s0} = 0.03\% \cdot 2\text{V} = 0.6\text{mV}$$
$$\text{Total errors}: \quad |A| \leq 0.04\text{mV} + 0.06\% \cdot U$$

The error curve may therefore only run outside the hatched areas of the diagram in Fig. 1.29.

A typical linearity error occurs during the quantization of an analog measurement signal. This will be discussed later.

- Random errors are measurement errors whose causes cannot be detected in detail and which cannot be repeated or predicted even under constant operating conditions. Random errors lead to scattering within a measurement series, and they cause comparable signal values to deviate from each other irregularly (stochastically). The measurement series can be a finite number of similar measured values of different measurement objects, it can also consist of temporally successive instantaneous values of the same measurement signal. The more random errors occur during a measurement, the lower

Fig. 1.29 Permissible and
impermissible ranges (*hatched*)
for the error curve of a
measuring device

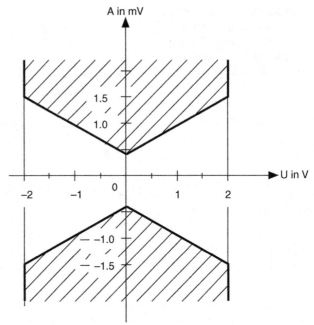

the precision of the measurement. Random errors can be recorded more reliably with
statistical methods the more individual values are available.

Figure 1.30 shows two ways of visualizing the scattering behavior of a measurement
series. In the upper figure, the individual measured values of a stress measurement series
are plotted as points on a scale beam. The lower figure shows a histogram in which the
entire variation range is divided into six equally wide intervals and in each interval, a vertical bar is shown, the height of which corresponds to the respective frequency density h'.
The frequency density for each interval is the ratio of the relative frequency h to the interval width ΔU. The relative frequency indicates the percentage of all measured values
within the considered interval. It is a measure of the probability that an individual signal
value falls within the respective interval. The frequency density h' is a measure for the
density of the measuring points in the upper figure. Figure 1.30 shows the scattering
behavior of a measurement series.

The larger the total number n of available signal or measurement values, the finer the
variation range can be subdivided and the finer the gradation of the histogram becomes.
For $n \to \infty$, the limiting case is a continuous curve, which is shown as a dashed line in the
figure below. In most cases, this curve or histogram can be described with sufficient accuracy by a symmetrical bell-shaped curve, the equation of which is then

$$h'_N = h'_{max} \cdot e^{-\frac{(U-\bar{U})^2}{2s_U^2}} \tag{1.10}$$

Fig. 1.30 Scattering behavior
of a measurement series. (**a**)
Representation of the individual measurement points on
a scale beam. (**b**) Histogram

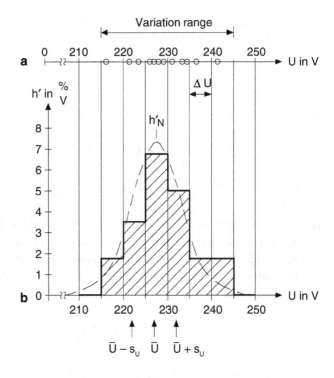

This is known as a normal distribution of the signal values.

If one wants to do without a detailed pictorial representation of the scattering behavior, one can limit oneself to the specification of the mean value and $\bar{U}$ the standard deviation s_U, which represent the statistical characteristic values (statistical signal parameters):

- The arithmetic mean of a $\bar{U}$ series of measurements is the most probable value within the range of variation. At has $U = \bar{U}$ the bell curve its maximum.
- The standard deviation s_U is a measure of the typical deviation of the individual values from the arithmetic mean. With a normal distribution, the inflection points of the bell curve are at the points "$\bar{U} - s_U$" and "$\bar{U} + s_U$". In this area, an average of 68.3% of all measured values is located. In each case, 15.9% of all individual values of a normally distributed measurement series are smaller than "$\bar{U} - s_U$" or larger than "$\bar{U} + s_U$". 95% of all individual values lie between the limits "$\bar{U} - 1.96s_U$" and "$\bar{U} + 1.96s_U$".

In general, the mean value $\bar{x}$ and the standard deviation s_x, can be calculated from n signal values $x_1, \ldots, x_n$ as follows:

$$\bar{x} = \frac{1}{n} \cdot \sum_{i=1}^{n} x_i \tag{1.11}$$

$$s_x = \sqrt{\frac{1}{n-1} \cdot \sum_{i=1}^{n} (x_i - \bar{x})^2} \tag{1.12}$$

For the statistical parameters of a continuous-time signal $x(t)$, $n \to \infty$ and the observation time T follows from (1.11)

$$\bar{x} = \frac{1}{T} \cdot \int_0^T x(t)\,dt \tag{1.13}$$

and that of the equation

$$s_x = \sqrt{\frac{1}{T} \cdot \int_0^T \left(x(t) - \bar{x}\right)^2 dt} = +\sqrt{X_{\text{eff}}^2 - \bar{x}^2} = X_{\sim\text{eff}} \tag{1.14}$$

The mean value of an alternating quantity is therefore its direct component, and the standard deviation is the effective value of the alternating component. The statistical parameters of a stochastic signal are the more reliable the less the internal dependencies of the signal are important, i.e. the longer the observation time T is.

Even in the case of random errors, it is useful to make relative specifications. For example, the coefficient of variation v_x is the ratio of the standard deviation s_x to the amount of the $|\bar{x}|$ mean value:

$$v_x = \frac{s_x}{|\bar{x}|} \tag{1.15}$$

For an alternating signal $x(t)$ (1.16)

$$v_x = \frac{s_x}{|\bar{x}|} = \frac{X_{\sim\text{eff}}}{|\bar{x}|} = W \tag{1.16}$$

The coefficient of variation v is identical to the ripple W of the signal.

With (1.11), (1.12), and (1.15) the result is calculated for the example in Fig. 1.30:

$$\bar{U} = 219\text{V}, \quad s_U = 7.0\text{V}, \quad v_U = 3.2\%$$

Even if no systematic errors are to be taken into account, the mean value is only the most probable, but not the true value x_w With the statement probability P (confidence level), this lies within an uncertainty range "$\bar{x} - u_z$" to "$\bar{x} - u_z$", which can be delimited around the mean value. The random component u_z of the measurement uncertainty is all the greater, the greater the selected confidence level P and the standard deviation s are and the smaller the number n of individual measured values is:

$$u_z = \frac{t_{(P,n)} \cdot s}{\sqrt{n}} \tag{1.17}$$

The same applies to the relative value:

$$\varepsilon_z = \frac{t_{(P,n)} \cdot v}{\sqrt{n}} \tag{1.18}$$

Fig. 1.31 Confidence factor t for normal distribution

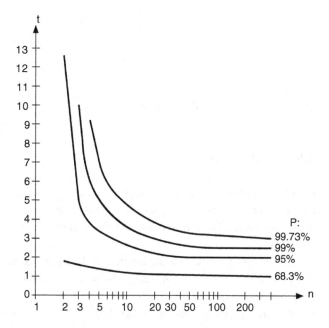

The confidence factor t in this equation is shown in Fig. 1.31 as a function of P and n. For $n \rightarrow \infty$, $u_z \rightarrow \infty$ assumes that the probable statement of the true value is sure to be consistent with the mean value.

In the example (Fig. 1.31) $n = 10$. With a chosen confidence level of $P = 99.7\%$, the results are $t = 4.2$ and $u_{zU} = 6.5$ V.

The systematic components u_s, ε_s ((1.8) and (1.9)) and the random components u_z, ε_z ((1.17) and (1.18)) of the absolute and relative measurement uncertainty, respectively, can be summarized by linear additions to resultant total values u and ε:

$$u = u_s + u_z \tag{1.19}$$

$$\varepsilon = \varepsilon_s + \varepsilon_z \tag{1.20}$$

A complete measurement result always contains information about the uncertainty, which may also lie within the specified number of digits. If the random component refers to a confidence level $P \neq 95\%$, this must also be specified.

In the example (Fig. 1.30), if the value $u_{sU} = 3.5$ V has been estimated for the systematic component on the basis of the measuring instrument properties, the total uncertainty of the measurement is

$$u_U = 6.5\text{V} + 3.5\text{V} = 10\text{V}$$

The complete measurement result is

$$
\left.
\begin{array}{ll}
U & = 219\,\mathrm{V} \qquad \pm 10\,\mathrm{V} \\
\quad\text{or} & \\
U & = 219\,\mathrm{V}(1 \quad \pm 4.6\%)
\end{array}
\right\} \text{with } P = 99\%
$$

$$\uparrow \qquad\qquad \uparrow \qquad\qquad\qquad \uparrow$$

Measured value   Uncertainty   Level of reliance

1.4.2   Sources of Error

In order to avoid, reduce or record measurement errors, their causes must be determined and the sources of error localized. In the following, a distinction is made between DUT errors, internal and external errors, and observation errors.

- DUT errors are errors that are inherent in the measured variable itself, e.g. if a DC voltage to be measured is superimposed by a noise voltage and the measured variable becomes uncertain.
- Measurement errors that depend exclusively on the characteristics of a measuring unit considered in isolation (e.g. of a single measuring element) shall be called internal errors or, in the case of a complete measuring instrument, instrument errors. Manufacturer's specifications, e.g. The error class of an indicating measuring instrument or calibration or error curves attached to the instrument always refers to the internal errors of the measuring unit.

Another example of an internal error is the rounding error of a PC. A digital calculator that squares a four-digit decimal number and rounds the result to four digits, in turn, produces an output signal of 1.789 for an input signal of 1.666. The systematic error of the output signal related to the correct value 1.789123 is therefore

$$
A_r = \frac{1.789 - 1.789123}{1.789123} = -6.87 \cdot 10^{-5}\,(\mathrm{m.v.})\,(\text{from Measured value})
$$

- When connecting several measuring elements to form a measuring chain or several measuring instruments to form a measuring system, it is usually not sufficient to superimpose the internal errors of all units according to the rules of error calculation, but it must also be considered that there are interactions between the elements of a measuring circuit. This results in additional external errors of the individual elements. External errors are not exclusively dependent on the properties of the considered measuring element itself, but they are also determined by the properties of the surrounding measuring circuit. These errors are therefore also referred to as circuit errors or process errors. Such errors cannot be specified by a manufacturer in a measuring instrument manual, since the manufacturer cannot know the individual test setup.

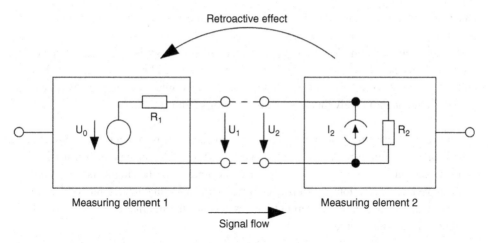

Fig. 1.32 Retroactive effect as a cause of external faults

Figure 1.32 shows a simple example where the output of a measuring element 1 and the input of a measuring element 2 are shown as equivalent circuits. Measuring element 1 generates a voltage U_0 as a measuring signal, which corresponds to its output voltage U_0 in no-load operation. If measuring element 2 is connected, there is a feedback effect on measuring element 1. The voltage U_1 changes under the influence of the load by measuring element 2, and the further processed voltage U_2 is not equal to U_0 but is equal to it:

$$U_2 = U_0 \cdot \frac{R_2}{R_1 + R_2} + I_2 \cdot \frac{R_1 \cdot R_2}{R_1 + R_2}$$ (1.21)

A systematic error occurs:

$$A = I_2 \cdot \frac{R_1 \cdot R_2}{R_1 + R_2} - U_0 \cdot \frac{R_1}{R_1 + R_2}$$ (1.22)

This error contains two parts. The first is positive and constant. This has the effect of a zero-point error. The second component is negative when the signal voltage U_0 is positive and depends on the signal voltage. If one of the resistors R_1 or R_2 is dependent on the measurement signal, which is often the case with electronic circuits, the feedback also causes a linearity error. From (1.21), it can also be seen that partial errors can either add up or compensate completely or partially, depending on the respective sign situation.

For each measurement, it must be taken into account that there is a reaction from the measuring device to the measured object. Whether the resulting errors are to be considered or negligible depends on the individual case. It is hardly conceivable that the surface temperature of the sun changes when measured from the earth with a radiation thermometer. On the other hand, it is only possible to measure the temperature of a single resistor in an integrated semiconductor circuit with a reasonable amount of effort and without feedback.

One task of the measurement technician is to design measuring equipment in such a way that its effects on the object to be measured are as small as possible.

An advantage of digital technology over analog technology is that external errors in digital circuits can always be avoided if certain regulations and recommendations for circuit design are observed. This means that even complex digital systems are always easier to understand quantitatively than comparable analog circuits.

- The person performing an experiment can also cause measurement errors, which are called observation errors. The parallax when reading a pointer-type measuring instrument can cause a systematic or random error, depending on the circumstances. Another example is the operation of a stopwatch. Due to the individual uncertainty of the actuation time, a measurement uncertainty arises because the internal error of the watch is almost always negligible.

1.4.3   Influence Errors

Measurement errors are often co-determined by influencing variables from which the respective measurement signal should theoretically be independent. Such influences are always undesirable, but rarely avoidable. The most important influencing variables in electronic measuring equipment are temperature, the frequency of the measured variable, magnetic or electric external fields, and time.

- Temperature influences: Most of the physical effects used in electronics are temperature-dependent and cause temperature drifts in almost all component parameters. These dependencies can usually only be determined empirically and are specified by the manufacturers as temperature coefficients or error limits. Some of the most important temperature effects for measurement electronics are listed below:
  - The charge carrier mobility in conductive materials decreases with increasing temperature and the charge carrier concentrations in semiconductors increase with temperature. This is the reason why metallic resistors usually have positive temperature coefficients (some ‰/K for pure metals) and semiconductor resistors usually have negative temperature coefficients (in the order of %/K). For particularly temperature-constant measuring resistors the material manganin is available, whose temperature coefficient is at most a few $10^{-5}/K$.
  - Diffusion stresses at boundary layers between different materials grow with temperature. The Seebeck effect is based on this, as shown in Fig. 1.33. At different temperatures ϑ_1 and ϑ_2 the contact points between the two materials A and B (are the diffusion stresses ΔU_1 and ΔU_2) are different. However, the voltages U_1 and U_2 before and after the contact point are different. This effect is particularly pronounced with semiconductor materials. In order to avoid measurement errors to a large extent,

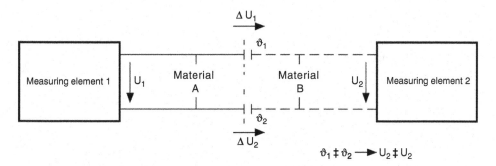

Fig. 1.33 Thermoelectric voltages as the cause of a DC voltage error on a measuring line

efforts are made to keep temperature gradients at critical points small and to select suitable material combinations. The combination of copper and manganese is particularly favorable, resulting in thermoelectric voltages of less than 1 µV per K temperature difference.

– The forward voltages of semiconductor diodes and base-emitter paths of transistors have negative temperature coefficients in the order of 1‰/K. Diodes with positive temperature coefficients can be used for temperature compensation.

– Reverse currents of diodes or transistor sections operated in the reverse direction are considerably temperature-dependent. Their temperature coefficients are in the order of 10%/K.

– The Z effect has a negative and the avalanche effect a positive temperature coefficient. Both effects are simultaneously effective in a Z-diode. The temperature coefficients almost compensate each other at Z voltages of about 5 V. Reference diodes that are temperature compensated with semiconductor diodes can have temperature coefficients of a few 10^{-6}/K and are used as temperature-constant reference voltage sources.

Effective measures to reduce temperature influences are the already mentioned temperature compensation by means of diodes or semiconductor resistors or a thermo-stabilization of the temperature-sensitive parts of the measuring device.

In electronic measuring circuits, unavoidable frequency dependencies must be expected:

• Cables have inductive reactances, have parasitic capacitances against each other and against the reference ground.

• Measuring resistors have reactive components which are caused by line capacitances and winding inductances.

• Capacitors and coils have frequency-dependent loss and quality factors.

• Diodes and blocked transistor paths have junction capacitances that act like capacitors connected in parallel.

• Base-emitter paths of bipolar transistors are capacitive due to the minority carriers stored in the base zones.

Fig. 1.34 Frequency error of
a first order linear low pass
filter. $A_r\nu$: relative magnitude
error. $A_r\varphi$: relative angular
error, related to one period

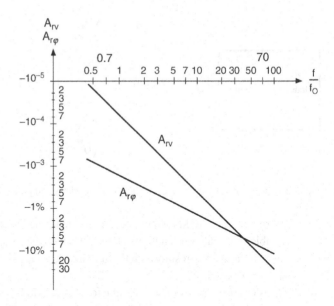

These and similar effects can lead to frequency-dependent measurement errors in both analog and digital measurement circuits. Analog measuring elements often behave like first-order linear low-pass filters. The relationship between an input alternating quantity and $\underline{X}_e$ an output alternating quantity is $\underline{X}_a$ then given by the relationship:

$$\underline{X}_a = \underline{X}_e \cdot \frac{\nu_0}{1+j(f/f_0)^2} = \underline{X}_e \cdot \underline{\nu} \tag{1.23}$$

where f_0 is the upper cut-off frequency. The complex factor ν is for $f = 0$ equal to the slope ν_0 of the static characteristic curve. For $f \neq 0$ an absolute value error of

$$A_{r\nu} = \frac{1}{\sqrt{1+j(f/f_0)^2}} - 1 \tag{1.24}$$

The angular error, which can be related to the angle 2π of a period, is calculated from

$$A_{r\varphi} = -\frac{1}{2\pi} \cdot \arctan(f/f_0) \tag{1.25}$$

These errors are shown in Fig. 1.34 as a function of the normalized frequency f/f_0. It can be seen that for $f/f_0 = 1$ the magnitude error is already about -30% and the angular error is over 10% of a period. The common practice of specifying the cut-off frequency for a measuring amplifier or an oscilloscope, e.g., is not very useful from a metrological point of view, since errors of this magnitude are not tolerable and must therefore no longer be measured at the cut-off frequency.

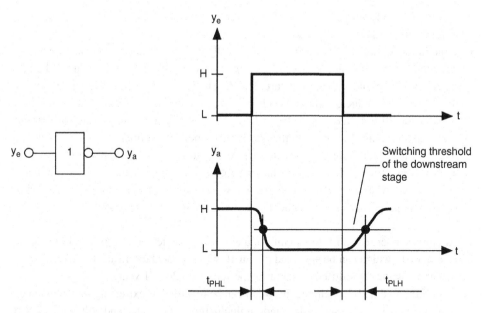

Fig. 1.35 Error due to the signal propagation times of a NOT-gate

Figure 1.34 shows that the permissible frequency limit for a precision measuring instrument may only be a few percent of the cut-off frequency.

In practice, the frequency behavior of a measuring system usually has to be examined empirically and, if necessary, optimized by suitable circuit measures.

In digital measuring elements, finite signal propagation times result from the causes described above. Figure 1.35 shows this using the example of a NON-gate. A positive signal jump from L potential to H potential at the input is delayed by the time t_{PLH} and becomes effective at the output. A negative signal jump from the H potential to the L potential at the input results in the delay time t_{PHL}. In general, these transit times are not the same in both switching directions, so that the effective duration of the output pulse is extended by the time with

$$\Delta t = t_{PLH} - t_{PHL}$$

This can mean a systematic measurement error when processing a time signal.

- External field influences: Electric and magnetic external fields can, if they change over time, cause errors in electronic measuring circuits due to influencing or inducing effects. The interference fields of technical alternating current (50 Hz), which are present almost everywhere, often have an extremely unpleasant effect, since their effects can hardly be distinguished from useful signals in terms of frequency. This can usually be remedied by magnetic or electrical shielding of the interference-sensitive circuit components. It must be considered, however, that this can result in additional circuit inductances and capacitances, which under certain circumstances can lead to a deterioration of the frequency response.

- Time influences: Data from electrical, especially electronic components are subject to temporal fluctuations and these drift apart more or less with time. This is due to the fact that the structure of a crystal structure, which is responsible for the conduction mechanism or a semiconductor effect, can change sporadically due to thermal lattice oscillations or radiation influences. These time influences are more pronounced in the first weeks and months after the manufacture of a component and then subside. This decay process can be accelerated by storing the device at an elevated temperature (artificial aging). Since time influences can never be completely eliminated in semiconductor components, wherever high precision is required, efforts are made to make measuring instrument properties as independent of semiconductor data as possible.

An important criterion for the quality of a measuring device or measuring circuit is the long-term drift, which can be specified e.g. in 10^4/Tag or ‰/Jahr. Instead, different error limits can be defined for different operating periods (e.g. 24 h, 1 year).

In addition to the drift phenomena, which can be regarded as extremely low-frequency, quasi-static change processes, faster random fluctuations of currents and voltages can also be observed in electronic circuits, which are referred to as noise. These noise signals generated in a circuit superimpose themselves on the measurement signals and falsify their instantaneous and peak values.

A measure for noise effects is the noise power P_r, which increases with the considered bandwidth Δf. Low-frequency noise in semiconductors is mainly due to the so-called spark noise, where the spectral noise power density dP_r/df to about 1 kHz decreases reciprocally to the frequency. At higher frequencies, resistance noise dominates and, in the case of bipolar transistors, current distribution noise dominates, both of which are virtually frequency independent.

Another time influence is the warm-up error, which is due to the temperature behavior of the measuring system. After switching on, a certain warm-up time is required, during which a stationary temperature state dependent on the self-heating of the components is reached. Only after reaching this final state are the internal errors of the measuring system constant. The warm-up time for precision measuring instruments can be up to 1 h.

1.4.4   Error Propagation

In general, a measurement result X according to a given function g depends on several independent measurement signals or circuit parameters $x_1, x_2, \ldots$

$$X = g\left(x_1, x_2, \ldots\right)$$

Fig. 1.36 Unloaded voltage divider

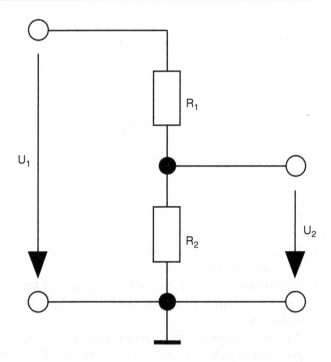

Then the systematic errors $A_1, A_2, \ldots$ also affect the independent variables $x_1, x_2, \ldots$ and the error A of the measurement result X. This effect is called error propagation. The following applies to absolute errors

$$A = \frac{\partial X}{\partial x_1} \cdot A_1 + \frac{\partial X}{\partial x_2} \cdot A_2 + \ldots \tag{1.26}$$

The following applies to relative errors

$$A_r = \frac{x_1}{X} \cdot \frac{\partial X}{\partial x_1} \cdot A_{r1} + \frac{x_2}{X} \cdot \frac{\partial X}{\partial x_2} \cdot A_{r2} + \ldots \tag{1.27}$$

Figure 1.36 shows an unloaded voltage divider consisting of the partial resistances R_1 and R_2 and is to be calculated. For the output voltage U_2 of the initially unloaded voltage divider ($R_L = \infty$) the following applies with the input voltage U_1:

$$U_2 = U_1 \cdot \frac{R_2}{R_1 + R_2} \tag{1.28}$$

According to the equation

$$A_{rU_2} = \frac{U_1}{U_2} \cdot \frac{\partial U_2}{\partial U_1} \cdot A_{rU_1} + \frac{R_1}{U_2} \cdot \frac{\partial U_2}{\partial R_1} \cdot A_{rR_1} + \frac{R_2}{U_2} \cdot \frac{\partial U_2}{\partial R_2} \cdot A_{rR_2} \tag{1.29}$$

For the three weighting factors in (1.29), by which the individual errors are to be multiplied before addition, the following applies

$$\frac{U_1}{U_2} \cdot \frac{\partial U_2}{\partial U_1} = 1 \tag{1.30}$$

$$\frac{R_1}{U_2} \cdot \frac{\partial U_2}{\partial R_1} = -\frac{R_1}{R_1 + R_2} \tag{1.31}$$

$$\frac{R_2}{U_2} \cdot \frac{\partial U_2}{\partial R_2} = -\frac{R_1}{R_1 + R_2} \tag{1.32}$$

It shows:

- A relative error of the A_{rU_1} input voltage U_1 causes an equally large error component of the output voltage U_2.
- Equal relative errors of the divider resistors have the same effect, but with opposite sign. For example, if both resistors are too large by 1%, the divider ratio remains unchanged.
- A relative resistance error of, for example, 1% contributes to the total error of the output voltage with less than 1%; the smaller the partial voltage, the greater the contribution.
- A load on the voltage divider by a load resistor R_L parallel to R_2 has the effect of reducing the divider resistance R_2. The error of R_2 is then for $R_L \gg R_2 / A_{rR_2} = -R_2 / R_L$. By (1.28) and (1.31) the following results

$$A_{rU_2} = -\frac{R_1 \cdot R_2}{(R_1 + R_2) R_L} \tag{1.33}$$

For $R_1 = 900\ \Omega$, $R_2 = 100\ \Omega$ and $R_L = 10\ k\Omega$, $A_{rU2} = -0.9\%$. The output voltage decreases by 0.9% due to the load. If the influence is to be less than 1‰, the load resistance according to (1.29) must be 900 times the value of R_2, i.e. greater than 90 kΩ.

The weighting factors from $\dfrac{x_1}{X} \cdot \dfrac{\partial X}{\partial x_1}$ (1.28) can be taken from the given function g without calculation if it is a product function of the form

$$X = x_1^a \cdot x_2^b \cdot x_3^c \cdot \ldots$$

...is trading. Then the rule is...

$$\frac{x_1}{X} \cdot \frac{\partial X}{\partial x_1} = a \qquad \frac{x_2}{X} \cdot \frac{\partial X}{\partial x_2} = b \qquad \frac{x_3}{X} \cdot \frac{\partial X}{\partial x_3} = c \tag{1.34}$$

If the systematic errors of the measurands $x_1, x_2, \ldots$ are only estimated and known as measurement uncertainties $u_1, u_2, \ldots$ (systematic components), then also for the result X only an uncertainty u concerning the worst case can be estimated. This case is characterized by the fact that all error components add up with the same sign. From (1.29) and (1.30) follows then:

$$u = \left| \frac{\partial X}{\partial x_1} \right| \cdot u_1 + \left| \frac{\partial X}{\partial x_2} \right| \cdot u_2 + \ldots \tag{1.35}$$

and

$$\varepsilon = \left| \frac{x_1}{X} \frac{\partial X}{\partial x_1} \right| \cdot \varepsilon_1 + \left| \frac{x_2}{X} \frac{\partial X}{\partial x_2} \right| \cdot \varepsilon_2 + \ldots \tag{1.36}$$

According to these equations, (1.30), (1.31) and (1.32) give the relative uncertainty of the ε_{U_2} output voltage U_2 of the voltage divider:

$$\varepsilon_{U_2} = \varepsilon_{U_1} + \frac{R_1}{(R_1 + R_2)} \left(\varepsilon_{R_1} + \varepsilon_{R_2} \right) \tag{1.37}$$

If the errors of the independent variables are random and independent of each other, their effects on the measurement result will be partly added and partly compensated. The standard deviation of the result is then obtained by adding the individual contributions geometrically:

$$s = \sqrt{ \left(\frac{\partial X}{\partial x_1} \right)^2 \cdot s_1^2 + \left(\frac{\partial X}{\partial x_2} \right)^2 \cdot s_2^2 + \ldots } \tag{1.38}$$

$$\varepsilon = \sqrt{ \left(\frac{x_1}{X} \cdot \frac{\partial X}{\partial x_1} \right)^2 \cdot \varepsilon_1^2 + \left(\frac{x_2}{X} \cdot \frac{\partial X}{\partial x_1} \right)^2 \cdot \varepsilon_2^2 + \ldots } \tag{1.39}$$

This is to be illustrated by an example, which at the same time shows that the relationship between the independent variables and the measurement result can not always be given as an equation, but also as a characteristic curve:

The transistor BC107/BC237, whose characteristic curve is shown in Fig. 1.37, is connected to $U_{CE} = 18$ V, which contains a ripple (50 Hz) from 1‰. The base current I_B is a direct current from 150 µA, superimposed by a noise component whose effective value is 750 nA. According to (1.16) the ripples are identical with the coefficients of variation ν of the quantities. Accordingly, the following applies accordingly (1.38)

Fig. 1.37 Characteristic output curve of transistor BC107/BC237 with operating point AP

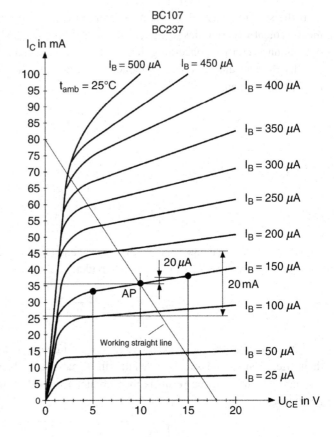

$$W_{I_C} = \sqrt{\left(\frac{U_{CE}}{I_C} \cdot \frac{\partial I_C}{\partial U_{CE}}\right)^2 \cdot W_{U_{CE}}^2 + \left(\frac{I_B}{I_C} \cdot \frac{\partial I_C}{\partial I_B}\right)^2 \cdot W_{I_B}^2} \qquad (1.40)$$

They are $W_{U_{CE}} = 1\text{‰}$ and $W_{I_B} = \dfrac{750\text{nA}}{150\mu A} = 0.5\text{‰}$.

The partial derivatives can be taken from the characteristic curve field:

$$\frac{\partial I_C}{\partial U_{CE}} \approx \frac{\Delta I_C}{\Delta U_{CE(I_B = -150\mu A)}} = \frac{20\text{mA}}{5\text{V}}$$

and

$$\frac{\partial I_C}{\partial I_B} \approx \frac{\Delta I_C}{\Delta I_{B(U_{CE} = 5V)}} = \frac{20\text{mA}}{100\mu A}$$

For the weighting factors, with which the ripples $W_{U_{CE}}$ and W_{I_B} to be multiplied, applies:

$$\frac{U_{CE}}{I_C} \cdot \frac{\partial I_C}{\partial U_{CE}} = \frac{10V \cdot 20mA}{36mA \cdot 5V} = 1.11$$

and

$$\frac{I_B}{I_C} \cdot \frac{\partial I_C}{\partial I_B} = \frac{150\mu A \cdot 20mA}{36mA \cdot 150\mu A} = 0.55$$

Eventually, the result is:

$$W_{I_C} = \sqrt{\left(1.11 \cdot 1\permil\right)^2 + \left(0.55 \cdot 0.5\permil\right)^2} = 1.17\permil$$

Here it is noticeable that the ripple of the resulting collector current is lower than that of each output variable alone, despite two overlapping effects.

1.4.5   Selection Criteria for Measuring Instruments

To be able to solve a measuring task optimally, it is necessary to select and combine the elements of the measuring device or the measuring instruments carefully according to technical and economic aspects.

- Measurement accuracy: The permissible systematic measurement error is specified by the task, and care must be taken to ensure that the total error of the measuring system resulting from all sources of error does not exceed this limit. On the other hand, it is uneconomical to increase the measuring accuracy significantly beyond the required level. Therefore the general rule in practice is
  Measure as accurately as possible, but not more accurately than necessary!
      Random errors do not affect the measuring accuracy if the measurements are taken sufficiently frequently or for a sufficiently long time so that the random influences are averaged out.
- Sensitivity: The sensitivity S of an analog measuring device or the transmission coefficient of an analogue measuring element is the ratio of effect to cause, i.e. the ratio of a change in the output variable x_a to the change in the input variable x_e causing it

$$S = \frac{\partial x_a}{\partial x_e} \approx \frac{\Delta x_a}{\Delta x_e} \tag{1.41}$$

The sensitivity of an analog indicating voltmeter is measured, for example, in mm/μV. The unit of the sensitivity of a resistance thermometer is Ω/K. The sensitivity can also be specified relatively by relating the relative changes of the output variable and the input variable to each other

$$S_r = \frac{\Delta x_a / x_a}{\Delta x_e / x_e} \tag{1.42}$$

High sensitivity is not synonymous with high accuracy. Often it is the other way round, because the more sensitive a measuring device is, the greater the measuring errors resulting from interference.

- Resolving power: Like the term "accuracy", resolving power can only be assessed qualitatively. The smaller the smallest measurand or measurement signal change that can still be detected, the greater the resolving power. With analog measuring elements, it is seldom possible to determine the smallest measurement quantity exactly, so that the specification of sensitivity is preferable. Conversely, the sensitivity of digital measuring elements does not make sense, since at least one of the quantities to be put into relation is a dimensionless code word. On the other hand, a measurement can be clearly specified here d. h. It corresponds to the quantization range ΔX and is referred to as a digital measuring step. Even more significant is the possible number of measuring steps. With digital devices, it is always finite, but often much larger than the number of actually distinguishable measuring steps of a comparable analog device.

If, for example, in the case of a five-digit resistance measuring instrument, the digital measuring step 10 is mΩ and all ten digits are permitted in each decimal place, the largest value that can be displayed is: 999.99 Ω, and the number of measuring steps is 100,000. However, with many measuring instruments, only either a "0" or a "1" can be displayed in the most significant decimal place (1 bit). In this case, the largest possible display is 199.99 Ω and the number of measuring steps is 20,000. This is called a 4 1/2-digit display.

The resolving power or the number of measuring steps of a measuring system has only indirectly something to do with its measuring accuracy. For example, the odometer in a motor vehicle can resolve a driving distance from 100 m. However, this does not mean that a total driving distance of, for example, 100,000 km is measured with an accuracy of +100 m, i.e. with an uncertainty of $10^{-6} = 1$! On the other hand, there is no point in equipping a digital voltmeter with a three-digit display only if its relative measurement uncertainty is less than 10^{-5} v. E. (of the measuring range end value).

- Measuring rate: It has already been explained that many analog and all digital signals can only change discontinuously, e.g. in a given clock. This cycle is often determined by the fact that in each measuring cycle the measured variable must be averaged over a sufficiently long measuring time to eliminate the influence of random errors. However, even a continuously changing analog signal at the output of a measuring element requires a finite adjustment time to adjust to the new signal value within specified error limits after a jump of the input signal.

- Each measuring system, therefore, has a finite measuring rate. It indicates how many independent individual measurements are possible one after another within a reference time interval. For manual evaluations, low measuring rates up to a maximum of 1 measurements/s are usually optimal, while for automatic measurement data processing in a PC, the highest possible measuring rates up to 10^6 measurements/s are aimed for.
- Frequency range: The most important frequency dependencies of electronic measuring equipment have already been discussed. The upper and lower frequency limits indicate the frequency range in which the respective measuring device operates within the specified error limits.

It is not technically possible, nor would it make economic sense, to optimize all the specified criteria simultaneously in a single device. An improvement of one of the features usually means a simultaneous reduction of at least one of the other features or an increase in equipment costs. In practice, therefore, a compromise must always be found between the desired benefit of the measuring system and the justifiable expense.

Components of the Electronic Data Acquisition

2

Summary

Since 1982, personal computers with the components microprocessor, RAM and ROM units, cache memory, hard disks, backup systems for data protection, and system peripherals have dominated many applications in electronics. However, when technicians, engineers, and scientists are dealing with electronic measurement technology, process control systems, digital control technology, analog controls with microcontrollers, converter systems, and other technical and scientific problems, operational amplifiers are used. Figure 2.1 shows the structure of a PC system for the acquisition of analog measured values, digital processing, and the output of analog values. In this setup, the operational amplifier is the most important component.

At the input of a measuring system, there is the sensor that converts a physical quantity into an analog voltage or current value. For a sensor to work, an operational amplifier is required in the sensor itself, which can function as an impedance converter, for example. The task of an impedance converter is to match between very high-impedance input impedance and low-impedance output impedance. The subsequent measuring amplifier must amplify the low voltage signal, e.g. in the voltage range, into a reasonable voltage signal, and in the voltage range. In practice, adjustable amplification factors are required for this purpose, which can be directly selected via a measurement processing software of the PC system. In practice it is often possible to choose between unipolar voltages from 0 V to ±10 V or bipolar voltages from ±5 V. While single-channel operation (measurement signal referenced to ground) dominates in electronic measurement technology, differential measurement is used in sensor technology.

To ensure optimum measurement operation in rough everyday use, passive or active low-pass filters are used, for example, to limit the bandwidth of the input signal. While passive filter circuits only require an ohmic resistor and a capacitor, active filter circuits have an operational amplifier with external circuitry. Active filter circuits can also be

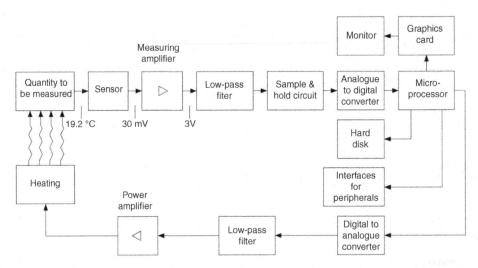

Fig. 2.1 Structure of a complete PC measurement system with AD conversion at the input, digital information processing, and final DA conversion for controlling external components

implemented with corresponding IC filter modules, which can be digitally programmed in wide ranges (cut-off frequency and quality) via a PC system. The calculation of active and passive filter circuits can be reduced to a minimum by using appropriate programs. The low-pass filter is followed by the so-called sample and hold (S&H) circuit, which temporarily stores the input signal. This provides a constant signal for the subsequent analog-to-digital converter for conversion.

With most AD converters (via 90%) the operational amplifier plays an important role because it controls the conversion. The AD converter generates a digital output value with a data format between 8 and 22 bit from an analog input signal. The converted data format is parallel, several circuit variants are possible, or serial. Via a PC system, the digital processing can now be carried out or the data (measurement data or already pre-processed information) can be stored and the information can be output on the monitor or printer. To access an external process, DA converters are required to convert the digital format into a corresponding analog output voltage. Here too, various operational amplifiers are required for conversion. The analog signal is then buffered at the output by a sample and hold circuit and fed to the controlled system via passive or active low-pass filters with a downstream power amplifier. The control loop is thus closed.

In many areas of process monitoring, there are low-frequency control processes that can be solved by software. Whereas in the past, PID controllers in analog circuit technology were equipped with transistors, there are now special PC programs that contain the algorithm for an integrated PID controller. Process monitoring and control tasks in the areas of temperature, speed, and torque monitoring, air conditioning, engine test benches, and pressure control can thus be performed more easily. The settings are no longer made in analog form by adjusting potentiometers, but digitally via the keyboard of a computer.

In general, a control system aims to maintain certain variables at specified setpoints. Disturbances which affect the process should have as little effect as possible on the variables to be controlled. A distinction is now made here between analog and digital control. In analog control systems, potentiometers and adjusters are used to enter the setpoint. With digital control systems, the setpoints can be adjusted via the PC keyboard.

Whereas with analog control algorithms the entire process is based on amplifiers and delay elements, with digital control all analog variables must first be converted into a digital format so that digital processing can be carried out. Using special auto sequences, extensive optimization strategies (e.g. according to Ziegler-Nichols, Hrones, and Reswick) can then be implemented, which facilitate the selection of the optimum control parameters. After the output variable has been calculated, it is output to the control loop via the DA converter or digital pulse converter PWM (pulse width modulator).

The advantage of a control system is the low expenditure on hardware and software because only one measurement is required if a disturbance variable is to be recorded and eliminated. The disadvantage is that occurring disturbances cannot be detected automatically, i.e. one measurement is required for each disturbance variable. Furthermore, all factors of the controlled system must be known to simulate the control system accordingly.

In a closed-loop control system, the disturbance variables are immediately detected and corrected by sensors. Another advantage is that a given value is maintained more accurately. The effort of a control system is many times higher and measurement is always necessary. A basic prerequisite for adapting or setting a controller to a controlled system is that precise information about the controlled system is available, i.e. without precise information about the system to be controlled, it is not possible to select or set a controller for this system.

Especially when working in connection with a PC or programmable logic controllers, digital control technology is used today. The mode of operation is very similar to the analog control loop. Like the analog control loop, the digital control loop is divided into a controller and a controlled system. With digital control technology, only the form of the signals that are transmitted is different.

Now the question arises on how a digital controller is structured and how it works. A digital controller, also known as a DDC controller (direct digital control), initially has an analog-to-digital converter (ADC) for the input variable, a digital comparator (DC), and a digital reference variable adjuster (WD). The control response (PID) is formed by a digital computer with a microprocessor or microcontroller, its digital output variable (YD) is converted into a continuous signal Y by a digital-to-analog converter (DAC) if a controller for continuous actuators is to be implemented. If a switching controller is required, the conversion of the output signal YD is simpler and only one digital pulse converter PWM (pulse width modulator) is required, which is often included in the computer and is realized by designing the computer program accordingly. The analog input variable X is converted in the AD converter into a binary coded signal which is processed accordingly on the computer.

2.1   Analog Amplifier Families

The quality of an application in analog electronics is determined by the specifications of the various types of amplifiers used. In practice, a distinction is made between the individual amplifier types, which are summarized in Table 2.1.

Other types of amplifiers play an essential role in signal processing for data acquisition.

Table 2.1 Designations for the individual operating modes of discrete, hybrid, and integrated amplifier circuits

Amplifier designations according to	
Purpose	Differential amplifier
	Operational amplifier
	Computing amplifier
	Universal amplifier
	Power amplifier
	Isolation amplifier
Performance	Preamplifier
	Amplifier
	Sound control amplifier
	Power amplifier
	Power amplifier
	Switching amplifier
Frequency	DC voltage amplifier
	AC voltage amplifier
	Pulse amplifier
	LF amplifier
	HF amplifier
	FM amplifier
Bandwidth	Broadband amplifier
	Selective amplifiers
	Frequency-dependent amplifiers
Coupling	DC-coupled
	AC coupled
	RC-coupled
	Transformer-coupled
	Isolation amplifier
Mode of operation	Single-ended amplifier
	Push-pull amplifier
	Complementary amplifiers
Operating mode	A-operation
	B-operation
	AB-operation
	Complementary operating modes

An isolation amplifier has the task of processing a small differential measurement signal, which can be superimposed on a high voltage level of several hundred or thousand volts. In principle, an isolation amplifier has the characteristics of an instrumentation amplifier, except that the common-mode voltage may be much higher. An instrumentation amplifier is a special operational amplifier with extensive features such as programmable gain, high input resistance, and internal band-limiting (low or bandpass).

For signal amplification in the microvolt range, the chopper amplifier is suitable. This type of amplifier causes only an extremely small input offset voltage. In general, chopper amplifiers can be used in the same applications as any standard operational amplifier, but concerning offset voltage, offset drift, and input current, significant improvements can be achieved by the chopped operation.

An electrometer amplifier has only a minimum input quiescent current of usually <1 pA. This amplifier converts a very small measuring current into a high voltage level that is easy to process. In practice, any impedance converter with a FET or MOSFET at its input is an electrometer amplifier. With the amplification of $v = 1$, there is a very high input resistance, the output has a standardized value of $R_a = 75\ \Omega$.

2.1.1   Internal Circuit Design of Operational Amplifiers

The integrated standard operational amplifier has been an essential component in many electronic systems for years. Until about 1970 it was common practice to develop an individual amplifier for each application. Since 1970, there has been an increasing tendency to concentrate the parts of the amplifier that provide the actual amplification in an integrated circuit and to achieve the especially desired characteristics by external circuitry. Such active circuits, which contain all components necessary for signal amplification, are called operational amplifiers. This designation comes from analog computing technology, in which these amplifiers were used on a large scale (operational amplifier = computing amplifier or OP for short). For more than 50 years, however, the applications have gone far beyond the scope of computing operations.

Using suitable external circuits of an operational amplifier, especially desired transmission characteristics can be achieved, which is the only way to make operational amplifiers universally applicable. To ensure that the characteristics of the connected amplifier depend only on the external circuitry, the operational amplifier circuit must meet various requirements. Therefore, an operational circuit has a rather complicated internal structure and consists of a large number of transistors, diodes, capacitors, and resistors.

Detailed knowledge of the internal structure of a surgical circuit and the internal functions is not necessary for the practitioner. However, the technician or engineer should know the internal functions so that he can optimally adapt the external circuitry to the internal workings. It is therefore not necessarily sufficient to regard the OR circuitry as a "black box". Next to the circuit symbol, there should be a technical background

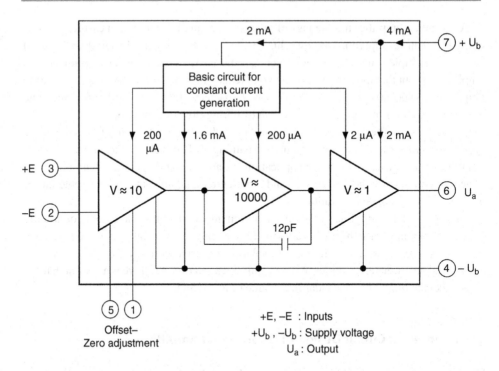

Fig. 2.2 Block diagram for the internal structure of a high-quality operational amplifier with individual functional units

knowledge for the individual type designations. To understand how a circuit built with standard operational amplifiers works, one must know the most important characteristics.

The internal block diagram of Fig. 2.2 shows that a high-quality operational amplifier consists of four functional groups:

- Input stage
- Voltage amplifier stage
- Power output stage
- Constant current sources

The number of integrated transistor output stages, diodes, and resistors vary greatly from one operational amplifier to another. Depending on the type of operational amplifier, the input stage contains either transistors or Darlington stages, FET or MOSFET transistors with resistance values up to $10^{25}\ \Omega$ to increase the input resistance. Each OP type has a different voltage amplifier stage with quite different and complex circuitry. There are two main types of power output stages, the circuit with a push-pull output stage or with an open collector.

The circuit symbol for an operational amplifier is shown in Fig. 2.3. As already shown in Fig. 2.2, the operational circuit has two inputs, one of which has an inverting and the other a non-inverting effect on the output. When a positive voltage is applied to the inverting input ($-E$), the output has a negative voltage value because the voltage is phase-shifted

Fig. 2.3 Circuit symbol, pin
assignment and pin connections for the operational
amplifier 741

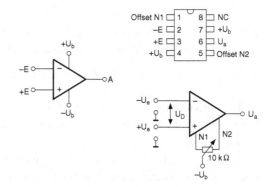

by 180°. There is no phase shift between the non-inverting input (+E) and the output.
Therefore, if a positive voltage is applied to the non-inverting input (+E), the output has a
positive voltage value.

Operational amplifiers are operated either with two balanced operating voltages of $+U_b$
and $-U_b$ if the output is equipped with a push-pull output stage, or with only one operating
voltage and ground if the device operates with a single-ended output, i.e. it has an output
with an open collector. In this case, an external working resistor is required.

As shown in Fig. 2.3, the voltages $+U_e$ and $-U_e$ at the inputs +E and −E, and the output
voltage U_a is referred to as a common reference point (ground). The adjuster can be used
to influence the symmetry of the operational amplifier 741.

2.1.2   Operating Modes of an Operational Amplifier

The ideal OP amplifies now since there is a differential amplifier on the input side according to the block diagram, only the differential voltage

$$U_D = +U_e - \left(-U_e\right)$$

with a gain factor designated V_0. The value V_0 is defined as no-load gain, differential gain,
or open-loop gain. The term "open-loop gain" does not mean that the output is unloaded.
The value V_0 is specified by the manufacturer and is approximately in the range of $2 \cdot 10^4$
to 10^5, depending on the type.

If the input voltages $+U_e$ and $-U_e$ and the open-circuit voltage gain V_0 *are* known, the
output voltage U_a can be calculated:

$$U_a = V_0 \cdot U_D = V_0 \left[+U_e - \left(-U_e\right)\right]$$

Two simplifying assumptions are now made to determine the designation of the inputs:
If the input −E is connected to ground, i.e. $-U_e = 0$ V, the output voltage is

$$U_a = V_0 \cdot U_D = V_0 \left[+U_e - \left(-U_e\right)\right] = V_0 \cdot \left(+U_e\right)$$

The output voltage U_a is therefore in phase with the input voltage $+U_e$. The input $+E$ is therefore described as non-inverting and is identified by a plus sign in the circuit symbol. If the input $+E$ is connected to the ground, i.e. $+U_e = 0$ V, then the output voltage

$$U_a = V_0 \cdot U_D = V_0 \left[+U_e - (-U_e) \right] = V_0 \cdot (-U_e)$$

The output voltage U_a is now in antiphase to the input voltage $-U_e$ at the inverting input. The input $-E$ is, therefore, called an inverting input and has a minus sign in the circuit symbol.

If the same voltage $+U_e = -U_e = U_{gl}$ is applied to the $+E$ and $-E$ input, $U_D = 0$ V. This operating mode is called common-mode control. According to $U_a = V_0 \cdot (-U_D)$, $U_a = 0$ should remain. However, this is not the case with the real operational amplifier and is referred to in this context as common-mode amplification

$$V_{gl} = \frac{U_a}{U_{gl}}$$

The value V_{gl} should at least be very small. The manufacturers specify the so-called common-mode rejection G in the data sheets:

$$G = \frac{V_0}{V_{gl}}$$

Typical values for G are 10^3 to 10^5.

The equation $U_a = V_0 \cdot U_D$ shows that the output voltage U_a (at constant V_0) rises or falls linearly with the differential input voltage. However, only until the value of the operating voltage is reached on the output side. A further increase of U_D then does not cause a change of U_a any more. The operational amplifier is overdriven, i.e. the output is in positive or negative saturation. The level of saturation depends on the operating voltage. Figure 2.4 shows these relationships for the transfer characteristic at the output of an operational amplifier.

Due to the high no-load gain of an operational amplifier, a small voltage difference at the two inputs is sufficient, since the amplifier is in the positive or negative saturation range. The internal open-circuit gain can be reduced by external feedback according to the circuit requirements. For this purpose, the user can choose between a frequency-dependent AC operation or a frequency-independent DC operation.

2.1.3   Transmission Characteristics of Operational Amplifiers

The possible applications of an operational amplifier can be divided into the following four general operating modes:

- Voltage amplifier
- Current amplifier

Fig. 2.4 Typical transfer characteristic of an operational amplifier

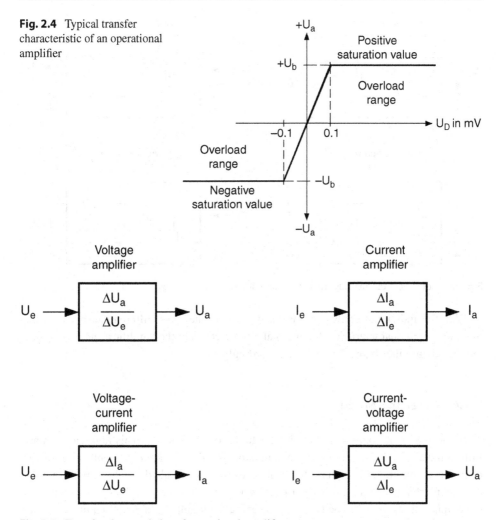

Fig. 2.5 Transfer characteristics of operational amplifiers

- Voltage-current amplifier (transconductance amplifier)
- Current-voltage amplifier (trans-impedance amplifier)

The criterion for the operating mode is the respective transfer characteristic between input and output. Figure 2.5 shows the typical characteristics of the different criteria.

Almost 99% of all analog applications are performed in practice with the voltage amplifier. The input voltage is amplified by a certain factor which is determined by the external circuitry of the OP module. The current amplifier only occurs in special applications in measurement technology, while the voltage-current amplifier and the current-voltage amplifier occupy a special position in analog circuit technology. The ideal characteristics and their equivalent circuit diagrams are shown in Fig. 2.6.

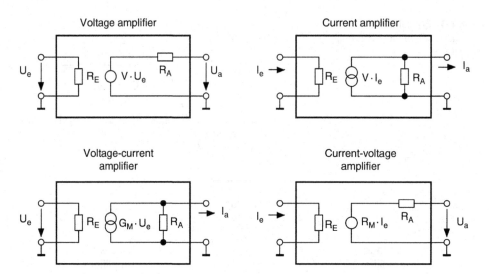

Fig. 2.6 Equivalent circuit diagrams of the individual amplifier types

With the individual equivalent circuit diagrams, one must differentiate between the input or output characteristics, the internal resistances, and the amplification factors. The individual amplifier types then behave accordingly.

2.1.4 Inverting Mode

Unconnected operational amplifiers have a high no-load gain V_0. This high gain is only fully utilized in a basic circuit, the comparator. In all other areas of application of the operational amplifier, the high no-load gain causes considerable difficulties. For example, a small interference voltage of only 0.1 mV at $V_0 = 5 \cdot 10^4$ could already cause a change in the output voltage of 5 V. The high gain of operational amplifiers is not required in practice, except for the comparators. The great advantage of the operational amplifier is that the amplification factor can be reduced to any desired value by a simple external circuit. With external circuitry, a distinction must be made between linear and non-linear components. The output function of operational amplifiers is largely dependent on these components with their specific characteristics.

In practice, the external circuitry is almost always designed as negative feedback. To be dependent only on the external circuitry, the operational amplifier must have ideal characteristics. This simplifies the approach considerably.

The operational amplifier has two inputs, the inverting and the non-inverting. The operating mode in positive and negative feedback can be realized very easily. If the output signal or a part of the output signal is fed back to the non-inverting input, the input signal and the fed-back signal are in phase and the feedback is obtained. If the output signal or

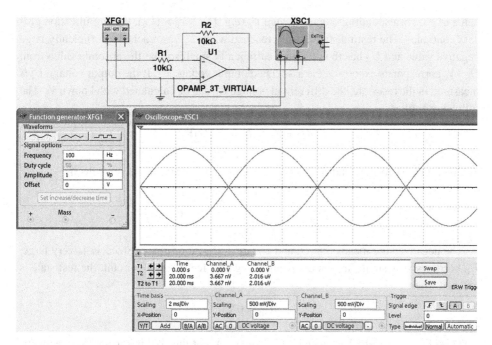

Fig. 2.7 Operational amplifier in inverting mode

part of the output signal is fed back to the inverting input, the input signal and the feedback signal are in the opposite phase and there is negative feedback.

In Fig. 2.7 a virtual operational amplifier is used. Multisim knows two types of operational amplifiers, virtual and real. Virtual operational amplifiers are divided into three and five-pole devices. The three-pin one already has two operating voltages and the five-pin one requires the operating voltage to be connected. The real operational amplifiers correspond to, e.g. the 741.

Via the function generator, a voltage from 1 V_s is applied to resistor R_1. The function generator is an ideal generator, so with $R_i = 0\ \Omega$. All setting values can be changed. Measurements are made with a two-channel oscilloscope and the two voltage curves show 1 V_s/100 Hz. The input and output voltage is 180° out of phase.

In practice, almost only negative feedback is used. Figure 2.7 shows an inverting operational amplifier with negative feedback. The mode of operation is easiest to understand when the sequence of a transient process is observed. If a positive DC voltage is applied to input E at a certain point in time, the output voltage does not change abruptly, i.e. without time delay, as is the case with all conventional amplifier circuits. At certain points in time, therefore, certain negative output voltage values will occur, which will affect the magnitude of U_D through negative feedback. For each voltage value U_a, there is therefore a resulting value for U_D. The more negative U_a becomes, the smaller U_D becomes.

The respective voltage U_D is, as is well known, amplified by the operational amplifier with the no-load amplification factor V_0. As long as the value $U_D \cdot V_0$ is greater than the value of the corresponding output voltage U_a (e.g. $U_{D2} \cdot V_0 > U_{a2}$), the amplification process continues. The resting state is only reached when U_a has reached a sufficiently large negative value and U_D has thus become sufficiently small so that the absolute value from $U_D \cdot V_0$ corresponds exactly to the associated output voltage U_a. If the output voltage U_a *is* measured in the rest state, the differential voltage U_D can be calculated with known V_0. The following applies:

$$U_a = V_0 \cdot \left(+U_e - \left(-U_e \right) \right)$$

Since $+U_e = 0$ V, the condition $U_D = -U_e$ applies to the circuit. This results in

$$U_a = -V_0 \cdot U_D \quad \text{or} \quad U_D = -\frac{U_a}{V_0}$$

The negative sign means that U_a and U_D are 180° out of phase. Since V_0 is very large, U_D becomes very small. In practice, with a negative feedback OP circuit, the rest state is reached when

$$U_D \approx 0\mathrm{V}$$

This idle state is called a "virtual" short circuit. According to this knowledge, the inverting amplifier can be easily calculated. With $U_D \approx 0$ V, the inverting input is connected to ground or 0 V, from which

$$U_{R1} = U_e \quad \text{and} \quad U_{R2} = -U_a$$

follows.

If you look at the series connection of R_1 and R_2, you can see that the current that flows through R_1 must also flow through R_2, because the input resistance of the operational amplifier is theoretically infinitely high. The two resistors R_1 and R_2 represent a voltage divider:

$$\frac{U_{R2}}{U_{R1}} = \frac{R_2}{R_1} \quad U_{R2} = U_{R1} \cdot \frac{R_2}{R_1}$$

With the relationships $U_{R1} = U_e$ and $U_{R2} = -U_a$ follows

$$-U_a = U_e \cdot \frac{R_2}{R_1} \quad U_a = -U_e \cdot \frac{R_2}{R_1}$$

This equation means:

(a) The input voltage is amplified with the factor $v = R_2/R_1$.
(b) The negative sign indicates that there is a phase shift of 180° between input and output voltage. This equation can also be written as

$$-U_a = U_e \cdot v \quad \text{or} \quad U_a = -U_e \cdot v$$

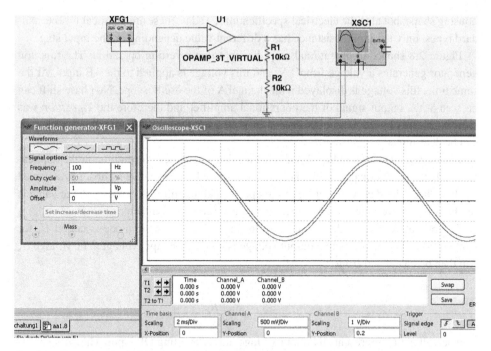

Fig. 2.8 Operational amplifier in non-inverting operation

where $v = R_2/R_1$ represents the gain in inverting OP mode. The gain of the inverting OP mode is therefore only dependent on the external circuitry. The choice of the resistance ratio can therefore be freely defined or set within wide limits independently of the open-circuit voltage gain.

2.1.5 Non-inverting Operation

When working in practice with non-inverting OP operation, one quickly moves from the standard types, such as the 741, to the special operational amplifiers with very high input resistance. For reasons of compatibility, all electrical specifications are identical except for the input resistance.

The non-inverting operation of an operational amplifier is called voltage-dependent voltage feedback.

The circuit of Fig. 2.8 shows a non-inverting amplifier operation. The input voltage is directly connected to the non-inverting input, which makes the high input resistance of the operational amplifier fully effective. If the standard operational amplifier 741 is used, an input resistance in the order of 500 kΩ is achieved. For the type ICL8007 with FET inputs, you can go to 10^{12} Ω, and if you use the ICL8500 with MOSFET inputs you can go to 1025 Ω. Important for these operational amplifiers are not only the pin assignment and the housing shape, but also the electrical specifications. Today, these are identical for the standard types, only the input resistance has a different value depending on the input stage.

Figure 2.8 shows an operational amplifier in a non-inverting operation. The function generator generates a voltage from 1 V_s and this voltage is applied to the +E input. At the same time, this voltage is displayed with channel A of the oscilloscope. No phase shift can be seen at the output signal of the operational amplifier and therefore the *Y-position* was shifted upwards by 0.2 V. Comparing the two curves in the oscilloscope, the gain of $v = 2$ can be determined.

For the circuit shown in Fig. 2.8 this means that in practice, depending on the application, only the operational amplifier needs to be replaced. The negative feedback is achieved by the voltage divider at the output of the operational amplifier. A certain voltage value is fed back by this voltage divider, which gives the circuit its gain factor.

The output voltage of the voltage divider in the non-inverting mode is called U_x for simplicity's sake, and the actual output voltage is used as the input voltage. The voltage U_x can then be calculated from

$$U_x = U_a \cdot \frac{R_2}{R_1 + R_2}$$

The voltage U_x represents the input voltage at the inverting OP input. The voltage difference between the two inputs is $U_D = 0$ V. Since $U_x = U_e$, the following applies

$$U_e = U_a \cdot \frac{R_2}{R_1 + R_2}$$

If you change this formula, you get

$$\frac{U_e}{U_a} = \frac{R_2}{R_1 + R_2}$$

The ratio U_e/U_a is not the amplification of a circuit, but the attenuation. To get the amplification $v = U_a/U_e$ you have to change the formula to U_a/U_e:

$$\frac{U_a}{U_e} = \frac{R_1 + R_2}{R_2} \quad \text{oder} \quad v = \frac{R_1 + R_2}{R_2}$$

A further transformation results in

$$v = \frac{R_1}{R_2} + \frac{R_2}{R_2} \quad \text{or} \quad v = \frac{R_1}{R_2} + 1 \quad \text{or} \quad v = 1 + \frac{R_1}{R_2}$$

Example: For the two resistors in the circuit select identical values from $R_1 = R_2 = 10$ kΩ. The amplification results in

$$v = 1 + \frac{R_1}{R_2} = 1 + \frac{10k\Omega}{10k\Omega} = 1 + 1 = 2$$

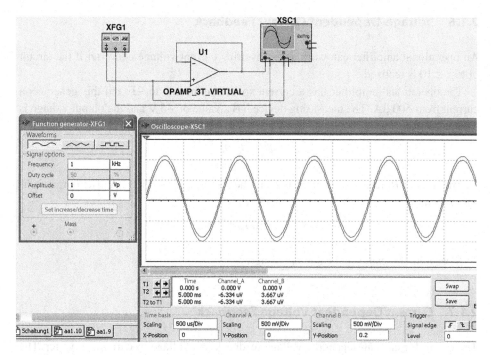

Fig. 2.9 Circuit of an electrometer amplifier

If both resistors have the same size, this results in practice in amplification of $v = 2$.

From this example you can now see the following: If we increase the resistor R_1 accordingly, the gain will be higher, because R_1 is in the counter of the fraction. If, on the other hand, we increase resistor R_2 accordingly, the gain decreases, since R_2 is in the denominator of the fraction.

In measuring practice, impedance converters are often required. Impedance converters are amplifiers with $v = 1$, which do not invert the input signal, have an extremely high input resistance, and an output resistance with the standard value of $R_a = 75\ \Omega$. These impedance converters are also known as electrometer amplifiers. This term was taken from tube technology.

Figure 2.9 shows the structure of an electrometer amplifier, where the resistor R_1 has a value of $0\ \Omega$ and R, is infinitely large. For this reason, the gain is $v = 1$. The full input resistance is used, while the output resistance is $75\ \Omega$.

Figure 2.9 shows an operational amplifier in non-inverting operation, as a so-called electrometer amplifier. The function generator generates a voltage from $1\ V_s$ and this voltage is applied to the +E input. At the same time, this voltage is displayed with channel A of the oscilloscope. No phase shift can be detected at the output signal of the operational amplifier and therefore the *Y-position* was shifted upwards by 0.2 V. Comparing the two curves in the oscilloscope, the gain of $v = 1$ can be determined. The characteristic of the electrometer amplifier is $v = 1$.

2.1.6   Voltage-Dependent Current Feedback

An operational amplifier can work as an inverting current-voltage converter if the circuit of Fig. 2.10 is realized.

The operational amplifier has a current source connected to −E and this generates a current from 500 μA. The measuring device has a value of −5 V and the output voltage is calculated from

$$-U_a = I_e \cdot R_1$$

The resistor R_1 has a value of 10 kΩ and the input current is 500 μA. The output voltage is then

$$-U_a = I_e \cdot R_1 = 500\mu A \cdot 10k\Omega = -5V$$

Measurement results and calculations agree.

2.1.7   Current Dependent Voltage Feedback

The circuit for current-dependent voltage feedback is a voltage-to-current converter. The load resistor R_L is neither directly connected to the ground nor to the operating voltage. Figure 2.11 shows the simulation of a voltage-to-current converter.

The input voltage is $U_e = 1$ V and the resistor R_1 has a value of 1 kΩ. The output current I_a of the operational amplifier is calculated from

$$I_a = \left(\frac{1}{R_1}\right) \cdot U_e$$

The circuit with the values is calculated from

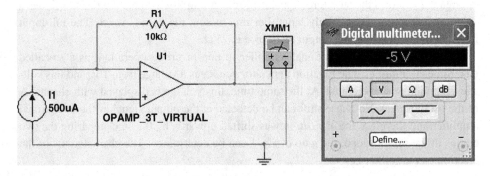

Fig. 2.10 Circuit of an inverting current-voltage converter

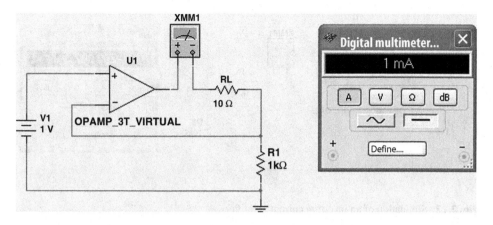

Fig. 2.11 Simulation of a voltage-to-current converter

$$I_a = \left(\frac{1}{1\text{k}\Omega}\right) \cdot 1\text{V} = 1\text{mA}$$

Measurement results and calculations agree.

2.1.8 Current Dependent Current Feedback

The circuit for current-dependent current feedback is an inverting current amplifier. The load resistor R_L is neither directly connected to the ground nor to the operating voltage. Figure 2.12 shows the simulation of a voltage-to-current converter.

The input current I_e is 100 µA and the two resistors R_1 and R_2 are each 10 kΩ. The output current is calculated by

$$I_a = -\left(1 + \frac{R_1}{R_2}\right) \cdot I_e$$

This results in the following value for the circuit of Fig. 2.12

$$I_a = -\left(1 + \frac{R_1}{R_2}\right) \cdot I_e = -\left(1 + \frac{10\text{k}\Omega}{10\text{k}\Omega}\right) \cdot 100\,\mu A = -200\,\mu A$$

Measurement results and calculations agree.

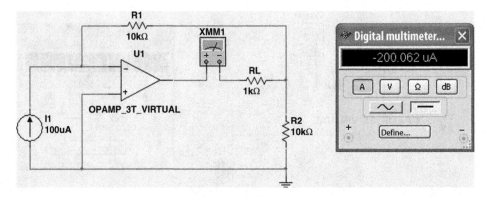

Fig. 2.12 Simulation of an inverting current amplifier

2.2   Linear and Non-linear Amplifier Circuits

Linear amplifier circuits are circuits in which the amplified output voltage is largely identical in amplitude form to the input voltage. In non-linear amplifier circuits, on the other hand, distortions occur which the user can consciously achieve by external circuitry.

2.2.1   Adder or Totalizer

If several resistors and voltage sources are present at the input of an operational amplifier, the function of an adder or totalizer is obtained. Figure 2.13 shows the circuit of an adder or totalizer.

As a result of the Kirchhoff rule, the virtual zero point of the circuit is valid:

$$I_{11} + I_{12} + I_2 = 0$$

If the equation is rearranged, the result is

$$I_{11} + I_{12} = -I_2$$

The currents are calculated from the voltages and resistances:

$$\frac{U_{e1}}{R_{11}} + \frac{U_{e2}}{R_{12}} = \frac{-U_a}{R_2}$$

If you use the same resistance values, the following applies

$$U_{e1} + U_{e2} = -U_a$$

The two input voltages are added and form a corresponding sum at the output. The input voltages can have positive and negative amplitudes.

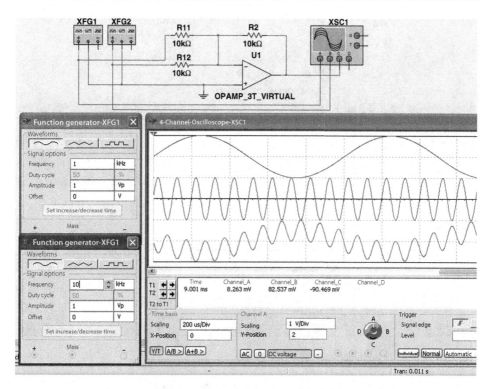

Fig. 2.13 Connection of an adder or totalizer

Example: The resistors in Fig. 2.13 should each have a value of 10 kΩ. The voltages have the following values: $U_{e\,1} = +3$ V and $U_{e\,2} = -4$ V. What is the output voltage?

$$U_{e1} + U_{e2} = -U_a$$
$$+3V + (-4V) = -1V$$

If the signs of the individual input voltages are taken into account, additions and subtractions can be carried out with a adder.

2.2.2 Operational Amplifier as an Integrator

As the circuit of Fig. 2.14 shows, integrators can be easily realized in combination with operational amplifiers. The input voltage is applied to resistor R, and the output voltage is fed back via capacitor C.

According to Kirchhoff's rule, the virtual zero of the circuit is valid for the virtual zero point of the circuit:

$$I_1 + I_2 = 0 \quad \text{or} \quad I_1 = -I_2$$

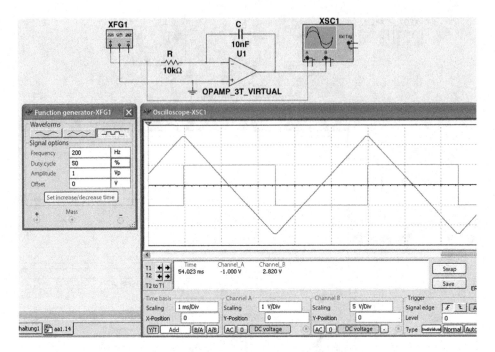

Fig. 2.14 Circuit of an integrator

This equation can be rewritten with the relations for I_1 and I_2:

$$I_1 = \frac{U_e}{R} \quad \text{and} \quad I_2 = \frac{dU_a}{dt} \cdot C$$

It follows from this

$$\frac{U_e}{R} = \frac{dU_a}{dt} \cdot C$$

If this equation is changed after the output voltage U_a, the result is

$$-U_a = \frac{1}{R \cdot C} \int_0^t U_e \, dt$$

This equation for the output voltage is correct if at the time t = 0 the output voltage has the value 0 V.

Example: What is the rate of change if you have an integrator with $R = 1$ MΩ and $C = 1$ µF and the input voltage is $U_e = 1$ V?

$$-U_a = \frac{1}{R \cdot C} \cdot U_e = \frac{1}{10^6\, \Omega \cdot 10^{-6}\, F} \cdot 1V = -1V / s$$

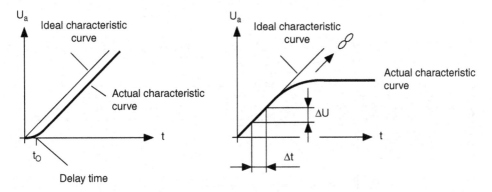

Fig. 2.15 Real and ideal conditions of integration with time delay at the output

At the output of the integrator, you have a rate of change from 1 V/s. This calculation assumes that the output voltage has $U_a = 0$ V. If the output voltage has a value U_0 at the time t = 0, the equation of the integrator has the following form:

$$-U_a = \frac{1}{R \cdot C} \int_0^t U_e dt + U_0$$

The voltage U_0 can have a positive or negative sign.

Normally an integrator works without delay. However, if the slew rate at the input is very high, it may take a certain short time until a signal appears at the output.

This delay time is shown in Fig. 2.15. The difference between the working range of the ideal and the actual characteristic curve is also shown. The ideal characteristic curve runs into the infinite range, but the actual one is limited by the operating voltage.

2.2.3   Differentiator with Operational Amplifier

While integrators can be used in analog computing technology, this is not the case with a differentiator because the real conditions are not fulfilled. The circuit in Fig. 2.16 shows a differentiator with a voltage diagram.

If the input voltage U_e is at 0 V, capacitor C is discharged and has a very low internal resistance. If the input voltage is suddenly changed, the output of the operational amplifier immediately goes into positive or negative saturation. In this example, a positive voltage jump occurs and the output of the operational amplifier goes into negative saturation.

Due to the negative output voltage, capacitor C can be charged via resistor R after an e-function. This charging changes the voltage on the capacitor and the output voltage on the operational amplifier goes back to 0 V. A negative voltage pulse, also called a "spike", is produced at the output of the operational amplifier. The output voltage can be calculated from

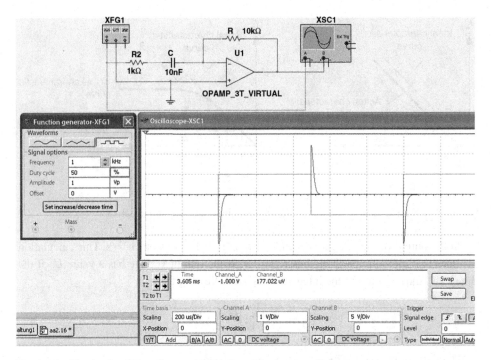

Fig. 2.16 Circuit of a differentiator with a voltage diagram

$$-U_a = R \cdot C \cdot \frac{dU_e}{dt}$$

The circuit of Fig. 2.16 is not the basic circuit, but a modified circuit. Normally the ideal circuit of a differentiator has only one capacitor and one resistor. The time constant for a real differentiator has a resistor in series with the capacitor as shown in Fig. 2.14. An improvement can be achieved by connecting a capacitor in series with resistor R. The calculation $\tau = R_2 \cdot C = R \cdot C_1$ and if $\omega \cdot \tau \ll 1$ it is $U_0 \approx -j\omega \cdot R \cdot C \cdot U_e$

$$u_0(t) = -R \cdot C \cdot \frac{du_e}{dt}$$

2.2.4 Differential Amplifier or Subtractor

Via an analog adder, input voltages with positive and negative signs can be added. In practice, however, this method has some disadvantages, which is why differential amplifiers or analog subtractors are used. A differential amplifier or subtracter is a combination of inverting and non-inverting amplifier operation. The corresponding circuit is shown in Fig. 2.17.

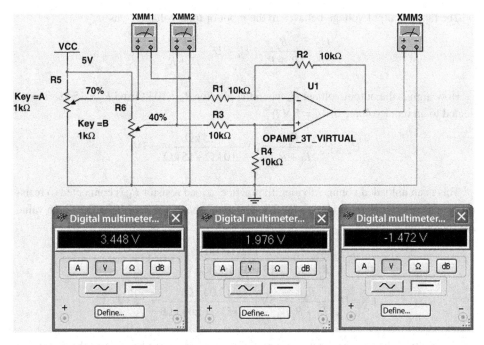

Fig. 2.17 Circuit of a differential amplifier or subtractor

As always, the output voltage is calculated based on an ideal operational amplifier, i.e. $U_D = 0$ V, because the differential voltage between the two inputs should be 0 V, and thus the inputs are at the same potential. The voltage at the +E input is calculated

$$U_{+E} = +U_e \frac{R_4}{R_3 + R_4}$$

The output voltage thus results from

$$U_a = \left(1 + \frac{R_2}{R_1}\right) \cdot \frac{R_4}{R_3 + R_4} \cdot \left(+U_e\right) - \frac{R_2}{R_1} \cdot \left(-U_e\right)$$

The calculation of the output voltage can be simplified if the condition $R_1 = R_3$ and $R_2 = R_4$ applies:

$$U_a = \frac{R_2}{R_1} \cdot \left(-U_e\right)$$

Two resistors R_1 and R_2 in series form a voltage divider. In the unloaded state the partial voltages behave like resistors:

$$\frac{U_1}{U_2} = \frac{R_1}{R_2}$$

The tapped output voltage behaves to the input or total voltage U_e as

$$\frac{U_a}{U_e} = \frac{R_2}{R_1 + R_2} \qquad U_a = U_e \cdot \frac{R_2}{R_1 + R_2}$$

How high is the output voltage U_a when the resistor $R_1 = 10\,\text{k}\Omega$ and $R_2 = 15\,\text{k}\Omega$ is connected to an input voltage of $U_e = 5\,\text{V}$?

$$U_a = U_e \cdot \frac{R_2}{R_1 + R_2} = 5\,\text{V} \cdot \frac{15\,\text{k}\Omega}{10\,\text{k}\Omega + 15\,\text{k}\Omega} = 3\,\text{V}$$

This is an unloaded voltage divider. In practice, a load resistor R_L is connected to resistor R_2. The resistor R_2 is parallel to R_L. The resistor R_L has a value of $100\,\text{k}\Omega$. Which value has the output voltage U_a?

$$R_2' = \frac{R_2 \cdot R_L}{R_2 + R_L} = \frac{15\,\text{k}\Omega \cdot 100\,\text{k}\Omega}{15\,\text{k}\Omega + 100\,\text{k}\Omega} = 13.04\,\text{k}\Omega$$

$$U_a = U_e \cdot \frac{R_2'}{R_1 + R_2'} = 5\,\text{V} \cdot \frac{13.04\,\text{k}\Omega}{10\,\text{k}\Omega + 13.04\,\text{k}\Omega} = 2.83\,\text{V}$$

Two voltage dividers connected in parallel form a resistance bridge and the resistance values behave like

$$\frac{R_1}{R_2} = \frac{R_3}{R_4}$$

On the left and right branches of the bridge, circuit are the same voltages d. h. Between these points, there is no voltage, the bridge is currentless and therefore balanced. In electronic measuring circuits, a bridge that is balanced in the rest state is often detuned by a changing bridge resistance value during the measurement. From the voltage change in the bridge diagonal one can then draw conclusions about the resistance change and it is valid:

$$\Delta U = U_1 - U_2$$

If the four resistors are chosen to have the same size, the voltages U_1 and U_2 are also equal and the differential voltage is $\Delta U = 0\,\text{V}$.

It's also important:

1. A bridge of linear resistors is not detuned when the operating voltage changes, because then the partial voltages change proportionally and the diagonal remains currentless.
2. A bridge is not detuned if two adjacent bridge branches are increased or decreased by the same factor. For example, if you choose 1.1 as a factor, then the equations

$$\frac{1.1 \cdot R_1}{1.1 \cdot R_2} = \frac{R_3}{R_4} \qquad \text{or} \qquad \frac{1.1 \cdot R_1}{R_2} = \frac{1.1 \cdot R_3}{R_4}$$

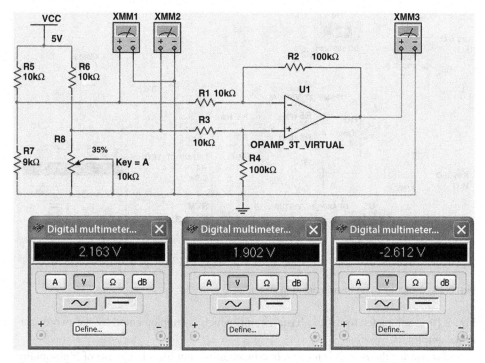

Fig. 2.18 Bridge voltage amplifier

if you change the adjacent resistors R_1 and R_2 or R_1 and R_3 in the same sense. You can see from the equations that the number factors stand out and the balance is maintained.

As Fig. 2.18 shows, a subtracter can be used as a bridge voltage amplifier. The measuring bridge consists of two voltage dividers that form the bridge circuit. The left voltage divider contains for example a photoresistor or an NTC resistor, and the output voltage for this bridge branch gives the voltage at the inverting input. The right voltage divider can be used to calibrate the measuring bridge.

The voltage amplification of the operational amplifier is $v = 10$. At a certain brightness at the photoresistor, the circuit is adjusted via the adjuster R_8 so that $U_a = 0$ V. If the brightness at the photoresistor increases, its internal resistance decreases and the voltage at the inverting input decreases. There is a difference between the two input voltages, which is amplified with $v = 10$, and the output voltage changes accordingly in a positive direction.

This bridge circuit can be found in the entire measurement technology. In one branch of the bridge circuit, the sensor is inserted, in the other branch, the adjustment possibility in the measuring circuit is available.

2.2.5   Instrumentation Amplifier

The circuit of a subtracter or differential amplifier can be improved considerably if this circuit is extended by two additional operational amplifiers. In the circuit shown in

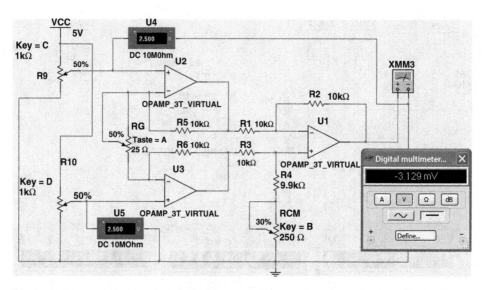

Fig. 2.19 Structure of an instrumentation amplifier

Fig. 2.19, a subtracter forms the basic circuit, which is supplemented by the two operational amplifiers.

Important for the realization of an instrumentation amplifier is the use of high-quality operational amplifiers. These are components with extremely low drift (0.1 µV/°C), low noise (<0.35 μV_{ss}), extremely high common-mode rejection (140 dB at $v = 1000$), and very low input currents (1 pA).

The input stage of the instrumentation amplifier requires high-quality amplifiers since offset, drift, and noise are multiplied by the set gain. The common-mode gain of the second stage is determined using the resistor RCM by applying an AC voltage with low frequency to the inputs U_{e1} and U_{e2} and adjusting the output voltage to zero. The resistors must have small tolerances to be able to use an adjustment adjuster with a small resistance value (drift!). Furthermore, they should have a low drift (R_1/R_2 and R_3/R_4) to allow a high common-mode rejection over the temperature range. The output voltage is calculated for the condition $R_1 = R_2 = R_3 = R_4 = R$ from

$$U_a = (U_{e1} - U_{e2}) \cdot \left(1 + \frac{2 \cdot R}{R_G}\right)$$

If the condition $R_1 = R_2 = R_3 = R_4 = R$ is not fulfilled, a further, undesired amplification occurs in the subtractor.

To suppress capacitive effects, the input lines should be actively shielded. With this measure, the signal lines and the shields are at the same potential. A current path for the input currents (bias-current) of the operational amplifiers must be ensured, otherwise, an impermissible operating state for the input stage will occur. For high amplification, the resistor RG becomes very low-impedance. Contact resistances have a direct influence on the amplification and drift. The taps to the inverting inputs should therefore be placed as close as possible to the resistor that determines the gain. With switchable amplifiers, spe-cial circuit measures may have to be taken. A low-impedance, the star-shaped ground connection is required to process small signals for resolutions with high dynamics. For large amplification, an adjustment of the input offset voltage is necessary.

2.2.6   Voltage and Current Measurement

The use of operational amplifiers can significantly improve the behavior of analog measur-ing devices. Figure 2.20 shows an amplifier circuit for a voltmeter. The operational ampli-fier is connected as an electrometer, i.e., one has a gain of $v = 1$, and the high input resistance of the operational amplifier is fully effective.

Normally an analog measuring instrument has a characteristic impedance of 10 kΩ/V. With the additional circuit shown in Fig. 2.20 the value can be increased to 10 MΩ/V. When the switch is in position 1, the voltage is directly connected to the non-inverting input. Since only very small input currents flow in the operational amplifiers, the voltage drop at the resistor of 10 kΩ is very small, and the measurement error is corre-spondingly low. By switching to positions 2 and 3, the voltage divider is tapped at certain points, which causes the input voltage to be divided down accordingly.

Important for the additional circuit are the two diodes 1N4001 (not in the simulation) between the inverting and non-inverting input. If a voltage greater than 0.6 V occurs, one of the two diodes is conductive, which means that the differential voltage cannot be greater than ±0.6 V.

The circuit of the ammeter shown in Fig. 2.21 consists of two parts: the current divider at the input and the operational amplifier, which is amplified by

$$v = 1 + \frac{9\,k\Omega}{1\,k\Omega + 9\,k\Omega} = 1.9$$

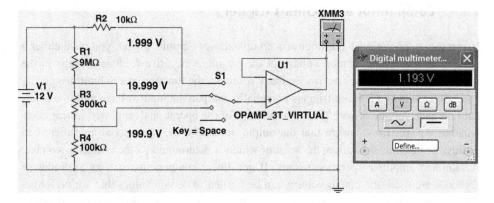

Fig. 2.20 Additional circuit for a voltmeter with three input ranges

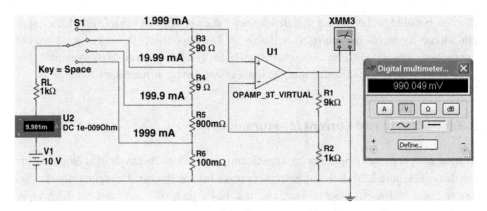

Fig. 2.21 Additional circuit for an ammeter with four measuring ranges

works. If the switch is in position 1, the voltage divider has a value of 100 Ω, and a current from 2 mA flows when the maximum input voltage is 2 V. Since the input voltage is applied directly to the non-inverting input of the operational amplifier, the result is a voltage of 3.8 V at the output of the operational amplifier.

Switching the switch to position 2 results in a resistance of 10 Ω between the switch and ground, and with an input voltage of 2 V, a current of 20 mA flows. At the output of the operational amplifier, a voltage of 3.8 V is produced. If the switch is switched to position 3, there is a resistance of 1 Ω to ground, and a current from 200 mA flows, which produces the maximum voltage of 3.8 V at the output of the operational amplifier. The maximum current flows when the circuit is in position 4. With an input voltage of 2 V, a current flows from 2 A to the ground, and at the output of the operational amplifier, a current of 3.8 V is measured. The output voltage of the operational amplifier is always dependent on the input voltage and thus indirectly on the current flowing to the ground via the switch.

2.3   Comparator and Schmitt Trigger

If you use an operational amplifier as a comparator or Schmitt trigger, you have either a positive or negative saturation voltage at the output. A negative feedback results in the function of a comparator, positive feedback results in the function of a Schmitt trigger.

Comparators and Schmitt triggers essentially compare an input voltage with the external reference voltage or with the output voltage of the operational amplifier. Linear components, e.g. resistors, ensure that the output voltage of the operational amplifier is in positive or negative saturation, the level of which is determined by the operating voltage. A saturated amplifier operation occurs. If non-linear components such as Si-diodes or Z-diodes are used, the output voltage can be limited to certain values that are no longer directly dependent on the operating voltage. In this case, this is called unsaturated amplifier operation.

2.3.1 Simple Voltage Comparator

In a simple voltage comparator, one essentially works with the saturated amplifier mode, i.e., the output voltage is in positive or negative saturation. The level of saturation depends only on the operating voltage.

The operational amplifier in the circuit shown in Fig. 2.22 has no external components to limit the no-load gain, which is therefore fully effective. If the two input voltages are different, the output of the operational amplifier is either in positive or negative saturation.

The circuit in Fig. 2.22 operates as a differential amplifier, whereby the no-load gain of the operational amplifier determines the switching behavior. The reference voltage U_{ref} determines the switching point of the operational amplifier. It applies:

$$U_{\mathrm{e}} > U_{\mathrm{ref}} \quad \rightarrow \quad U_{\mathrm{a}} = +U_{\mathrm{sätt}} \approx +U_{\mathrm{b}}$$
$$U_{\mathrm{e}} < U_{\mathrm{ref}} \quad \rightarrow \quad U_{\mathrm{a}} = -U_{\mathrm{sätt}} \approx -U_{\mathrm{b}}$$

If the input voltage is higher than the reference voltage, the output is in positive saturation because the input voltage is at the non-inverting input. If this connection scheme is changed, the behavior of the circuit changes.

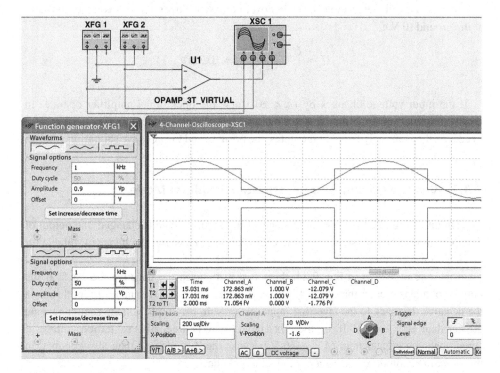

Fig. 2.22 Operational amplifier for comparing the input voltage U_{e} (function generator 1) with the reference voltage U_{ref} (function generator 2)

A problem with these circuits is the differential voltage between the two inputs. If the differential voltage becomes too large, the OP input stage may be destroyed. In practice, you have two input resistors, followed by two diodes in an antiparallel circuit. If the voltage difference between the two inputs is greater than +0.6 V one of the two diodes becomes conductive and limits the difference voltage accordingly. This results in overvoltage protection for the operational amplifier.

The simplest comparator circuit is shown in Fig. 2.23. At the inverting input, there is the voltage U_e, which is compared with the reference voltage U_{ref}. Since this circuit works without negative feedback, the full no-load gain is effective. The following applies to the output voltage

$$U_e < U_{ref} \quad \rightarrow \quad U_a = U_{a_{max}}$$
$$U_e > U_{ref} \quad \rightarrow \quad U_a = U_{a_{min}}$$

If the input voltage U_e is lower than the reference voltage U_{ref}, the output of the operational amplifier is in positive saturation, i.e., the output has $U_{a_{max}}$. If the input voltage U_e is greater than U_{ref}, the output voltage is in negative saturation, i.e. at $U_{a_{min}}$. The reference voltage can be continuously adjusted with the potentiometer at the non-inverting input, allowing the user to determine the reference voltage according to his requirements.

The no-load gain determines the switchover point when the reference voltage is exactly at the ground (0 V):

$$U_e = \frac{U_a}{v} = \frac{\pm 15 V}{50,000} = \pm 0.3 mV$$

If the input voltage changes by $U_e < \pm 0.3$ mV, the operational amplifier operates in analog amplifier mode. If the input voltage exceeds this value, the output voltage is digitized, i.e., there are only two voltage states, the positive and the negative saturation voltage.

2.3.2   Voltage Comparator in Saturated Amplifier Mode

The circuit of Fig. 2.23 has a disadvantage: If the input voltage is changed, the output of the operational amplifier switches to either $+U_b$ or $-U_b$. Thus, this circuit is not suitable for driving digital circuits.

In the circuit shown in Fig. 2.24, there is a Z-diode in the negative feedback. If the output voltage is greater than $U_Z = 4.7$ V, the Z-diode is conductive and a current flows via the negative feedback to the inverting input. As a result, the output voltage cannot become greater than $U_a = +4.7$ V. If, on the other hand, there is a negative saturation voltage, it cannot become greater than $U_a = -0.6$ V because the Z-diode will then also conduct. In this case, the Z-diode works as a normal silicon diode.

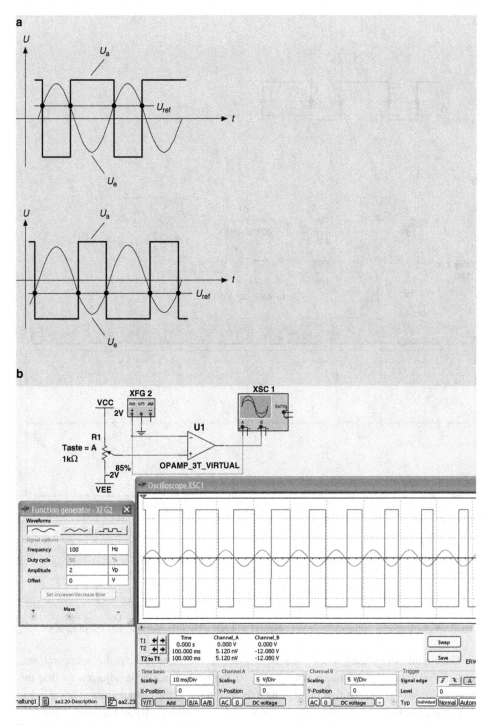

Fig. 2.23 Simple comparator circuit

a

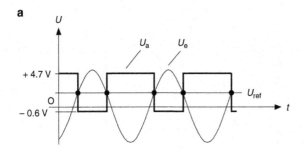

b

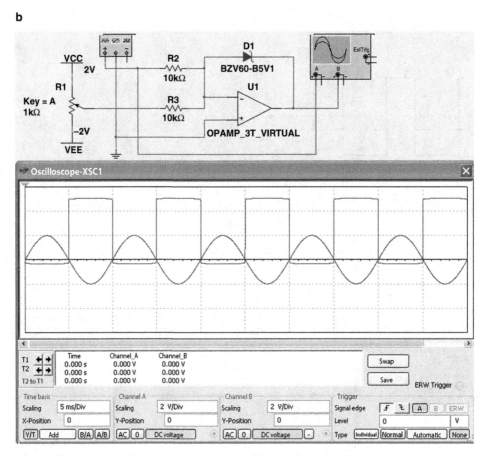

Fig. 2.24 Circuit of a comparator for the control of digital circuits with the voltage diagram

The circuit of Fig. 2.24 is designed as an adder. At one input there is the voltage U_e and at the other input the reference voltage U_{ref}. If the potentiometer is adjusted so that the input is $U_{ref} = 0$ V, the positive input voltage is inverted and the output voltage has a negative value, but this is limited by the Z-diode at $U_{a_{min}} = -0.6$ V. If, on the other hand, the input voltage is less than 0 V, the positive input voltage is inverted and the output voltage has a positive value, which is limited by the Z-diode at $U_{a_{max}} = +4.7\text{V}$. It applies to the output voltage:

$$U_e < U_{ref} \quad \rightarrow \quad U_a = U_{a_{max}}$$
$$U_e > U_{ref} \quad \rightarrow \quad U_a = U_{a_{min}}$$

With the potentiometer the reference voltage U_{ref} between +12 V and −12 V can be changed at will, i.e., the switching threshold between the positive (+4.7 V) and the negative (−0.6 V) output voltage is only determined by the reference voltage.

Many applications in electronics require a zero voltage comparator, as shown in Fig. 2.25. However, this simple circuit works very effectively. Due to the non-linear characteristic of the Z-diode, the output of the operational amplifier immediately goes into the positive and negative saturation voltage, which is then limited by the Z-diode to +4.7 V and −0.6 V. The input voltage U_e acts on the inverting OP input, while the non-inverting OP input is connected to ground. Any voltage difference between these two inputs brings the output voltage to +4.7 V or −0.6 V.

$$U_e < 0\text{V} \quad \rightarrow \quad U_a = -0.6\text{V}$$
$$U_e > 0\text{V} \quad \rightarrow \quad U_a = +4.7\text{V}$$

By selecting the Z-diode, the level of the output voltage in positive and negative direction can be determined. Connecting two Z-diodes or one Z-diode in series with a normal diode results in corresponding output voltages.

In the circuit shown in Fig. 2.26, there are two Z-diodes in the negative feedback, which are connected in series. If, for example, diode D_1 is conductive, the diffusion voltage of diode D_2 must be added to the Z-diode voltage of D_1. It applies:

$$U_e > 0 \quad \rightarrow \quad U_a = \left(U_{Z1} + U_{D2}\right)$$
$$U_e < 0 \quad \rightarrow \quad U_a = \left(U_{Z2} + U_{D1}\right)$$

The ohmic value of resistor R_1 is not important but should be chosen in practice with $R_1 = 10\text{ k}\Omega$.

The input voltage of this circuit is directly connected to the non-inverting input of the operational amplifier, which results in very high input resistance of the circuit.

2.3.3   Window Comparator

The parallel connection of two comparators results in a window comparator, the circuit of which is shown in Fig. 2.27.

a

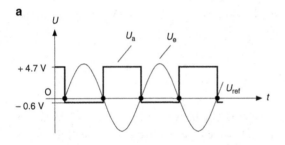

b

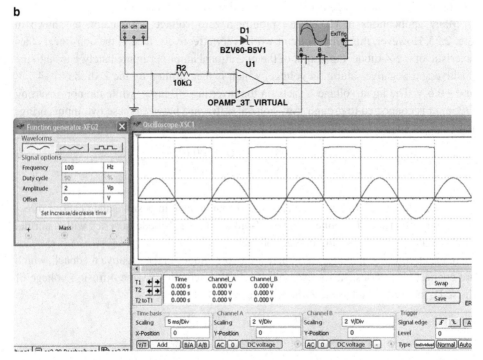

Fig. 2.25 Circuit of a zero voltage comparator with a voltage diagram

The input voltage U_e is compared with the lower and upper reference voltage via the two operational amplifiers. The conditions for the reference voltages must be observed so that the changeover points for the voltage diagram are valid. It applies:

$$U_{ref1} \leq U_e \leq U_{ref2}$$

The input voltage U_e is connected to the non-inverting input of the upper and the inverting input of the lower operational amplifier. The reference voltages U_1 and U_2 are connected to the two operational amplifiers accordingly. The two operational outputs control an AND gate, and here the digital link is made to generate an output signal. The circuit has a 1 signal at the output if the input conditions of the two reference voltages are met.

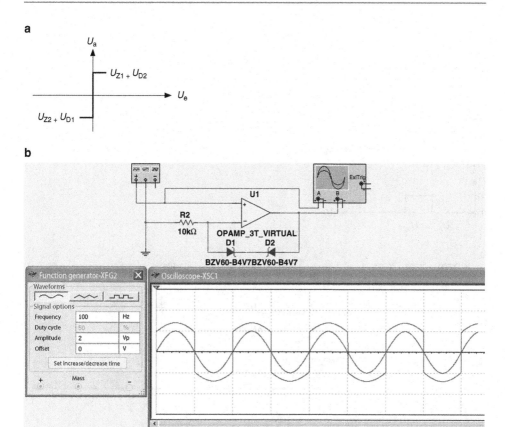

Fig. 2.26 Circuit and voltage diagram for a zero voltage detector

2.3.4 Three-Point Comparator

Normally, in practice only one comparator with two output values is needed, the two-point comparator. For some applications in control engineering, however, three output values are required. This switching of a three-step comparator is achieved by inserting a diode bridge in the negative feedback line, as shown in Fig. 2.28.

The two reference voltage terminals $\pm U_{\text{ref}}$ can be used to determine the threshold values S of the comparator. These values are calculated from

$$S_1 = \frac{R_1}{R_2} \cdot \left(-U_{\text{ref}} + U_{\text{D}}\right) - U_{\text{ref1}}$$

$$S_2 = \frac{R_1}{R_2} \cdot \left(+U_{\text{ref}} + U_{\text{D}}\right) - U_{\text{ref1}}$$

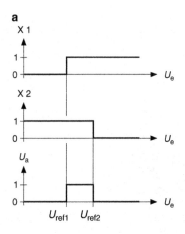

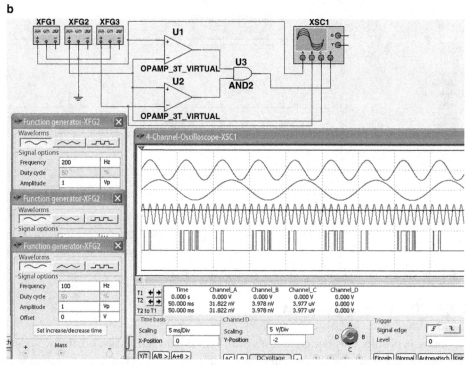

Fig. 2.27   Circuit and voltage diagram for a window comparator

The two threshold values depend on the reference voltages $+U_{ref}$ and $-U_{ref}$ applied. In addition to these values, there is the forward voltage U_D of the diodes with 0.6 V and the reference voltage $U_{ref\,1}$, which is parallel to the input voltage. Important with this circuit are the same values of the commonly marked resistors.

a

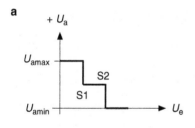

b

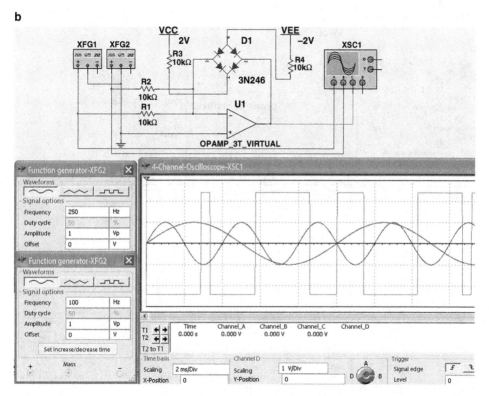

Fig. 2.28 Implementation of a three-point comparator with a characteristic curve

2.3.5   Schmitt Trigger

Whereas the comparator does not require a switching hysteresis, the Schmitt trigger generates a hysteresis using circuitry measures, which has numerous advantages in practice. In the basic circuits of the Schmitt trigger, a distinction is made between the inverting or non-inverting operating mode and a saturated or non-saturated behavior.

In the circuit shown in Fig. 2.29, the input voltage U_e is directly connected to the inverting input of the operational amplifier, while part of the output voltage U_a is coupled to the non-inverting input via the voltage divider. If the input voltage U_e has a negative value, the

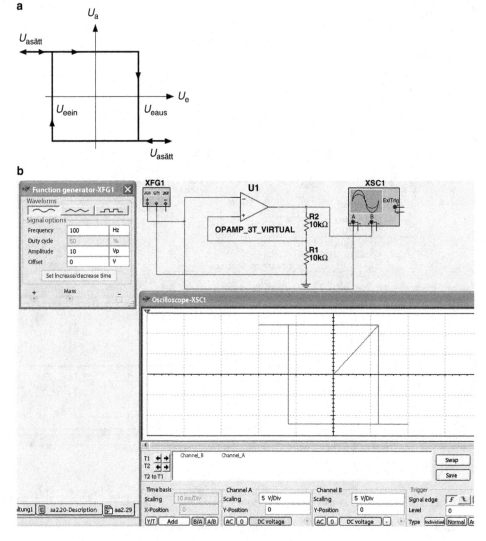

Fig. 2.29 Circuit and transfer characteristic of a Schmitt trigger

output voltage U_a is in the positive saturation voltage. A part of the positive output voltage is coupled $U_{a_{sätt}}$ via the voltage divider to the non-inverting input. The voltage at the non-inverting input of the operational amplifier is calculated from

$$U_x = \frac{R_1}{R_1 + R_2} \cdot +U_{a_{sätt}}$$

The condition at the output only changes when the input voltage exceeds a certain positive value. If the input voltage is at a negative value and is increased, the positive output voltage remains stable in its saturation. Only when the input voltage U_e becomes more positive than the reference voltage at the other input U_e, the output U_a switches over and is in negative saturation. Due to the voltage divider, the non-inverting input now has a negative voltage, which is

$$U_x = \frac{R_1}{R_1 + R_2} \cdot -U_{a_{sätt}}$$

The output voltage remains in negative saturation until the input voltage is again more negative than the voltage U_x. The two changeover points determine the hysteresis, which can be calculated as follows:

$$U_H = \frac{R_1}{R_1 + R_2} \cdot \left[+U_{a_{sätt}} - \left(-U_{a_{sätt}} \right) \right]$$

In the circuit shown in Fig. 2.30, the voltage at the non-inverting input is compared with the ground (0 V). In principle, one has a zero point Schmitt trigger.

If the input voltage U_e of the Schmitt trigger is at a negative potential, the output of the operational amplifier is in its negative saturation state, and via the voltage divider, a part of the output voltage is at the non-inverting input of the operational amplifier. This voltage is calculated from

$$U_x = \frac{R_1}{R_2} \cdot -U_{a_{sätt}}$$

If the input voltage increases, the voltage U_x also becomes more positive. If this voltage exceeds the 0 V limit, the output of the operational amplifier switches to positive saturation. This also changes the sign for the voltage U_x. If the input voltage decreases, the output remains stable in its saturation, and the voltage U_x is calculated from

$$U_x = \frac{R_1}{R_2} \cdot +U_{a_{sätt}}$$

By reducing the input voltage, the voltage U_x becomes negative. If this voltage falls below the 0 V limit of the inverting input, the operational amplifier switches over, and the output voltage is again in negative saturation. This type of positive feedback results in a switching hysteresis which can be calculated from

$$U_H = \frac{R_1}{R_2} \cdot \left[+U_{a_{sätt}} - \left(-U_{a_{sätt}} \right) \right]$$

In the circuit of Fig. 2.31, the input voltage U_e is at the inverting input. In contrast, the reference voltage U_{ref} drives the voltage divider, which consists of the two resistors R_1 and R_2. The center of the voltage divider is connected to the non-inverting input of the

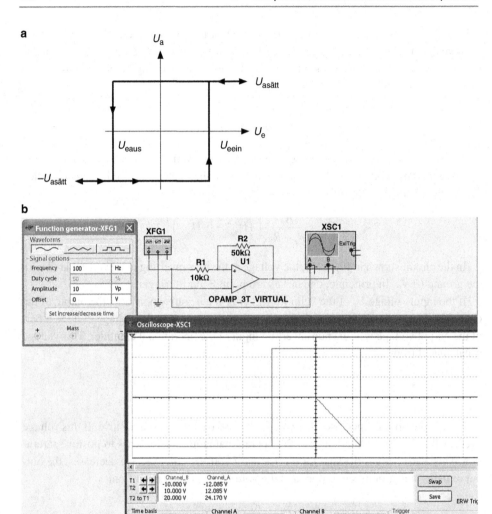

Fig. 2.30 Circuit and transfer characteristic of a Schmitt trigger with non-inverting output

operational amplifier. The magnitude of this voltage depends on the reference voltage and
the output voltage.

The two threshold values are calculated:

$$S_1 = U_{\text{ref}} - U_{\text{a}_{\text{sätt}}} \cdot \left(\frac{R_1}{R_1 + R_2} \right)$$

$$S_2 = U_{\text{ref}} + U_{\text{a}_{\text{sätt}}} \cdot \left(\frac{R_1}{R_1 + R_2} \right)$$

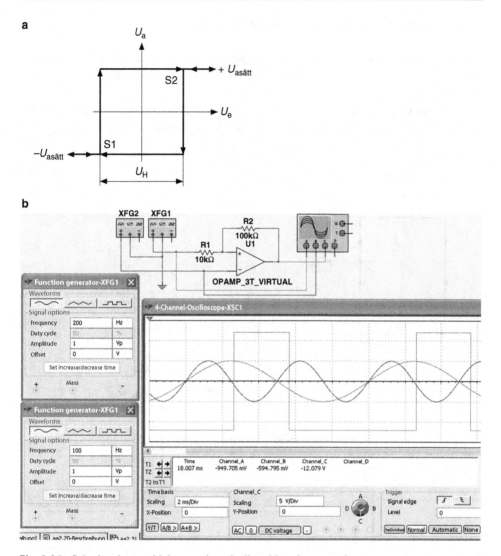

Fig. 2.31 Schmitt trigger with hysteresis and adjustable reference voltage

In the two equations, the negative saturation voltage must be used for the threshold value S_1 and the positive saturation voltage for S_2. In this case, a value of ± 12 V is used for an operating voltage of ± 11 V, since there is no corresponding limitation in the negative feedback. The calculation of the hysteresis results in

$$U_H = U_{ref} + \left(U_{a_{sätt}} - U_{ref}\right) \cdot \left(\frac{R_1}{R_1 + R_2}\right)$$

2.3.6   Schmitt Trigger in Non-saturated Mode

Linear components can be used to operate a Schmitt trigger in saturated mode. If non-linear components such as diodes or Z-diodes are used, the result is a non-saturated mode of operation.

A Schmitt trigger circuit that operates in non-saturated mode is shown in Fig. 2.32. In the negative feedback, there are two Z-diodes which are operated against each other. The

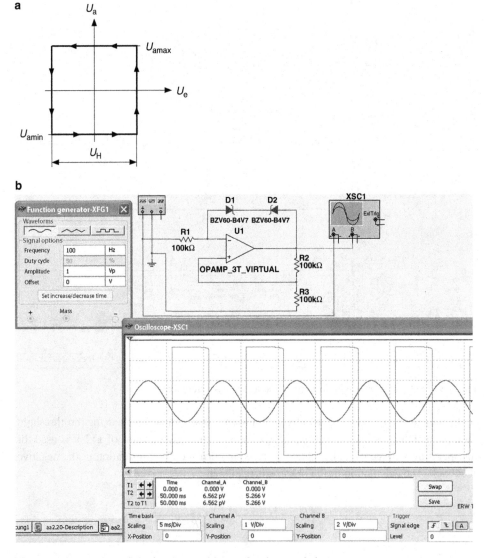

Fig. 2.32 Zero voltage Schmitt trigger with transfer characteristic

range of the two Z-diodes is ±5.3 V, because to each Z-voltage one has to add the lock voltage of 0.6 V of the normal silicon diodes.

The comparison voltage or the reference voltage of the circuit is generated by the voltage divider at the output. A part of the output voltage is coupled to the non-inverting input. The two threshold values for the switch-on and switch-off range are calculated as follows

$$\text{Switch -on level}: \quad U_{on} = \left(\frac{R_3}{R_2 + R_3} \right) \cdot \left(+U_{a_{sätt}} \right)$$

$$\text{Switch -off level}: \quad U_{off} = \left(\frac{R_3}{R_2 + R_3} \right) \cdot \left(-U_{a_{sätt}} \right)$$

As and $-U_{a_{sätt}}$ you have to $+U_{a_{sätt}}$ insert the voltage at the two Z-diodes with $U_g = 5.3$ V. The switching hysteresis U_H is then calculated from

$$U_H = U_g \cdot \frac{R_3}{R_2}$$

The output voltage of the circuit is

$$U_a = U_g \cdot \left(\frac{R_3}{R_2 + R_3} \right)$$

In the circuit shown in Fig. 2.33, the linear coupling is used via the two resistors R_1 and R_2, while the two diodes and the resistor R_3 form a non-linear voltage divider. The resistor R_3 can be calculated from

$$R_3 = \frac{R_1 \cdot R_2}{R_1 + R_2}$$

Since the difference voltage between the two inputs of an operational amplifier is $U_D = 0$ V, the two changeover points can be calculated:

$$U_e > +U_H \quad \rightarrow \quad +U_a = +U_e + U_S \cdot \frac{R_1 + R_2}{R_2}$$

$$U_e < +U_H \quad \rightarrow \quad -U_a = -U_e + U_S \cdot \frac{R_1 + R_2}{R_2}$$

The voltage drop $_{US}$ at the diodes is indicated with $U_D \approx 0.7$ V. With a positive output voltage the upper diode is conductive, with a negative value the lower diode is conductive.

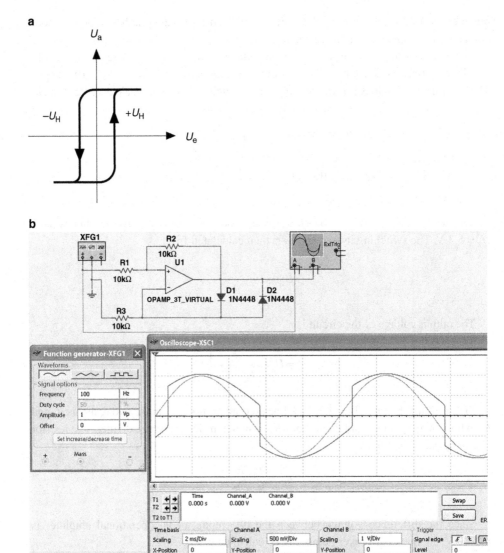

Fig. 2.33 Schmitt trigger in the non-saturated mode

2.3.7 Comparator with Tilting Behavior

Often noise or other interference signals are superimposed on the input signal. If the input voltage changes only slowly over time near the changeover point, these amplified interference signals may cause multiple undefined tipping to and fro. You can avoid this indeterminacy of the comparator output signal by feeding back the comparator.

If the feedback is positive (positive feedback) and the loop gain is greater than one, a circuit with flip-flop behavior is created. Because of the DC coupling, the circuit will tilt even if the input voltage rises or falls as slowly as desired.

In practice, you have two basic circuits. Analogous to the two basic circuits with an operational amplifier, two basic circuits can be distinguished in a coupled operational amplifier or comparator circuit depending on whether the input signal is fed to the inverting or the non-inverting input: the inverting and the non-inverting basic circuit. The static transfer characteristic curve is similar to a hysteresis curve. As a result of positive feedback, an upper and a lower threshold value occur and the difference is the switching hysteresis. Input voltages that are smaller than this hysteresis cannot tilt the Schmitt trigger back and forth, but can at most tilt it into a position where it remains. As we will now see, the hysteresis cannot be selected to be arbitrarily small, otherwise, the tipping condition of the circuit is no longer fulfilled. It is advisable to select a slightly larger hysteresis than the amplitude of the interference voltage superimposed on the input voltage. Then the trigger circuit is not switched back and forth unintentionally by the interference voltage.

If you compare the structure of a Schmitt trigger in a non-inverting operation, you can see that there is voltage feedforward. Combining these two relationships yields the transfer function of the circuit

$$\frac{U_a}{U_e} = \frac{v}{1 - k \cdot v}$$

If the loop gain $k \cdot v$ is equal to or greater than one, self-excitation occurs, i.e., the circuit acts as an oscillator (toggle circuit). Even the noise of the circuit itself causes the output voltage to oscillate almost abruptly due to the very high loop gain when the comparator is in its active operating range. After the flip-flop process is initiated, the comparator runs into positive or negative saturation and remains there.

To ensure that the described feedback process is as fast as possible and with a high degree of reliability, the Schmitt trigger circuit is dimensioned in practice so that $k \cdot v \gg 1$ applies. A certain increase in the loop gain during the tipping over process can be achieved by switching on a small capacitor C'. It acts as a dynamic reduction of resistor R_2 and thus increases the feedback factor k during the overturning process. The time constant C' $(R_2 \| R_1)$ should be chosen in the order of the rise or fall time of the output voltage $u_a(t)$. It is unfavorable to choose a much larger time constant since undesired transients can be the result.

A further improvement of the dynamic behavior can be achieved, the overdriving of the comparator or operational amplifier can be avoided by non-linear feedback. It has a negative effect on the circuit.

If the loop gain $k \cdot v$ has dropped to the value one, the switching hysteresis becomes zero. This means that it only occurs if the circuit exhibits tipping behavior. If the loop gain is sufficiently high, it is independent of the gain v. Practical experience clearly shows that arbitrarily small hysteresis values cannot be achieved, because the feedback factor k must

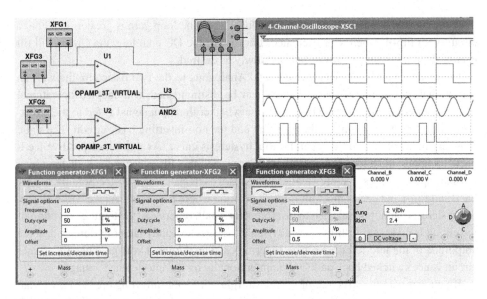

Fig. 2.34 Precision Schmitt trigger with RS flip-flop

only be reduced to such an extent that $k \cdot v > 1$ is still valid to ensure a sufficiently safe and fast tipping process.

The precision Schmitt trigger with RS flip-flop of Fig. 2.34 shows a very precise comparator circuit with hysteresis adjustable within wide limits, which can be realized e.g. by using a double operational amplifier. If the input voltage U_e exceeds the upper threshold U_2, the flip-flop consisting of the NAND gates G_1 and G_2 tilts to the position $U_e = $ H. The circuit does not flip back until U_e falls below the lower threshold U_1.

2.4 Measuring Bridges

A distinction is made between DC and AC measuring bridges. With direct current measuring bridges, a distinction is made between unloaded and loaded voltage dividers and bridge circuits. These measuring bridges only use ohmic resistors. In contrast to the AC measuring bridges, ohmic resistors, capacitors, and coils in a bridge circuit are operated with AC voltage.

2.4.1 Unloaded Voltage Divider

The unloaded voltage divider is a series connection of resistors, as shown in the circuit of Fig. 2.35.

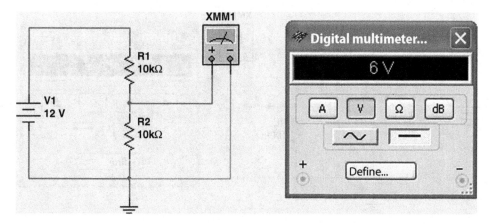

Fig. 2.35 Circuit of an unloaded voltage divider

In Fig. 2.35a DC voltage source with $U_e = 12$ V is available as the input voltage. The output voltage U_a is calculated from

$$U_a = U_e \cdot \frac{R_2}{R_1 + R_2} = 12\,\text{V} \cdot \frac{10\,\text{k}\Omega}{10\,\text{k}\Omega + 10\,\text{k}\Omega} = 6\,\text{V}$$

The ratio of R_1 to R_2 determines the output voltage.

Measurement results and calculations are identical.

2.4.2 Loaded Voltage Divider

The voltage divider is loaded by a load resistor R_L and the current and voltage ratios change, as the circuit of Fig. 2.36 shows.

The load resistor R_L is parallel to the resistor R_2 and this parallel connection is calculated from

$$R_2 \parallel R_L = \frac{R_2 \cdot R_L}{R_2 + R_L} = \frac{10\,\text{k}\Omega \cdot 100\,\text{k}\Omega}{10\,\text{k}\Omega + 100\,\text{k}\Omega} = 9.09\,\text{k}\Omega$$

The output voltage for a loaded voltage divider is obtained from

$$U_a = U_e \cdot \frac{R_2 \parallel R_L}{R_1 + R_2 \parallel R_L} = 12\,\text{V} \cdot \frac{9.09\,\text{k}\Omega}{10\,\text{k}\Omega + 9.09\,\text{k}\Omega} = 5.714\,\text{V}$$

Measurement results and calculations are identical.

Figure 2.37 shows voltages and currents on the loaded voltage divider. The deviation from the linearity of the unloaded voltage divider increases as the load resistance R_L decreases. The calculation is

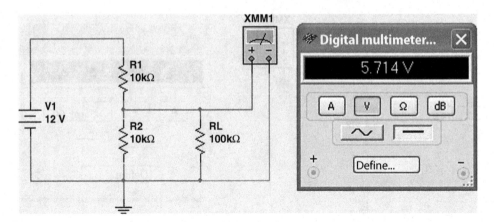

Fig. 2.36 Circuit of a loaded voltage divider

$$\frac{U_a}{U_e} = \frac{R_2 \parallel R_L}{R_1 + R_2 \parallel R_L} = \frac{a}{1 + \left(\dfrac{R}{R_L}\right)\left(a - a^2\right)} = \frac{I_L}{I_{L_{max}}} \quad \text{with } a = R_2 / R$$

The value of resistor R is the total resistance of R_1 and R_2.

2.4.3 Bridge Circuit

The bridge circuit consists of the parallel connection of two voltage dividers. If the voltage divider R_1 and R_2 divide the voltage applied between points A and B in the same ratio as the voltage divider of R_3 and R_4, there is no voltage between points C and D and the bridge is balanced.

The resistance measuring bridge is also called Wheatstone measuring bridge and it is used for direct resistance comparison. The accuracy achieved with it is determined by the accuracy of the reference standards and the sensitivity of the zero indicators (measuring instrument) in the diagonal branch of the bridge. The sensitivity shall be selected in such a way that the smallest change in the normal resistance causes a still detectable change in the indication of the zero indicators. The sensitivity of the zero indicators must therefore increase with increasing accuracy of the measuring bridge. The highest accuracy of technical measuring bridges is 10^{-5}, their measuring range is between 1 Ω and 1 MΩ, the error limit is 0.02% and they are suitable for DC and AC operation.

If in Fig. 2.38 the resistor R_3 has a value of 10 kΩ, the condition

$$\frac{R_1}{R_2} = \frac{R_3}{R_4}$$

and the bridge is aligned.

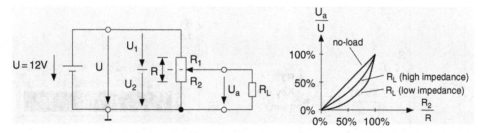

Fig. 2.37 Voltages and currents at the loaded voltage divider

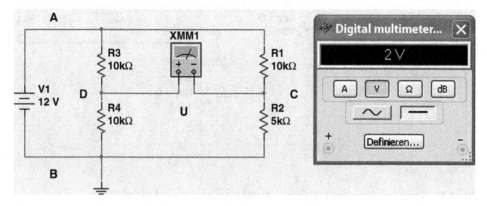

Fig. 2.38 Circuit for a bridge circuit (Wheatstone measuring bridge)

The voltmeter indicates a voltage of U = 2 V. The right voltage divider has an output voltage of

$$U_C = U_e \cdot \frac{R_2}{R_1 + R_2} = 12\,V \cdot \frac{5\,k\Omega}{10\,k\Omega + 5\,k\Omega} = 4\,V$$

The output voltage of the left voltage divider has $U_D = 6$ V and the bridge or differential voltage is

$$\Delta U = U_D - U_C = 6\,V - 4\,V = 2\,V$$

Measurement results and calculations are identical.

For resistance measurements in the range from 1 μΩ to 10 Ω the Thomson measuring bridge is used. The error limit is 0.1%. The direct tapping at the resistors reduces the influence of the line resistances and the contact resistances are compensated. Figure 2.39 shows the Thomson measuring bridge.

For the measurement of very small amounts of resistance, the double bridge according to Thomson is used. The resistances of R_3 and R_4, one of which represents the test object,

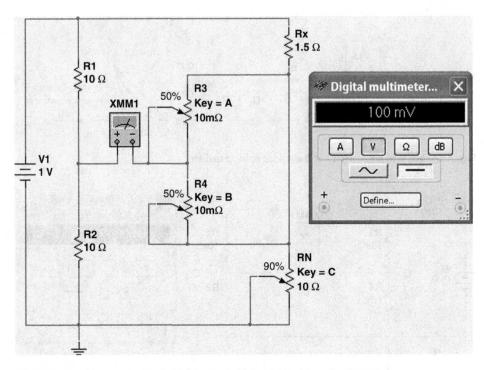

Fig. 2.39 Thomson measuring bridge for low ohmic resistors from 1 μΩ to 10 Ω

the other a normal resistance of the same order of magnitude, are equipped with potential terminals. If in this arrangement $R_1/R_2 = R_3/R_4 = R_x/R_N$, the adjustment condition

$$R_x = \frac{R_N \cdot R_1}{R_2} = \frac{R_N \cdot R_3}{R_4}$$

The measurement is therefore unaffected by lead and contact resistances. The simultaneous fulfillment of the adjustment conditions and the additional conditions is only possible with the aid of "double resistors" because a change of R_2 also requires a change of R_4.

2.4.4 Simple Capacitance Measuring Bridge

AC bridges can be used to determine resistances, inductances, capacitances, loss angles, frequencies, distortion factors, etc. For AC voltage, the adjustment condition applies analogously to the complex resistances of the $\underline{Z}$ bridge. With $\underline{Z} = R + jX$ the equation is divided into two conditions

$$R_2 \cdot R_3 - X_2 \cdot X_3 = R_1 \cdot R_4 - X_1 \cdot X_4$$
$$R_2 \cdot X_3 + R_3 \cdot X_2 = R_1 \cdot X_4 + R_4 \cdot X_1$$

Fig. 2.40 Circuit of a general
measuring bridge

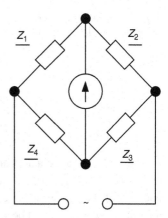

This results in three possibilities for the bridge adjustment:

1. Modification of two bridge elements
2. Modification of a bridge element
3. Adjustment not possible

In case 1, the bridge adjustment can converge or diverge, depending on the choice of adjustment elements. The divergence of the bridge and thus the practical impracticability of the adjustment occurs when the influence of each adjustment element equally affects both components of the complex diagonal voltage. Here the resistance values of both adjustment elements must be included in both adjustment conditions.

In case 2, one of the two adjustment conditions is always fulfilled.

Case 3 occurs when no operating case can be established which brings the diagonal voltage to zero. A pseudo-calibration can thus be established, which is characterized by the fact that the diagonal voltage as measured variable and the bridge input voltage as reference variables are perpendicular to each other.

A sinusoidal alternating voltage is required for the alternating voltage measuring bridges. Figure 2.40 shows the circuit of a general measuring bridge.

The matching conditions apply

$$\underline{Z}_1 \cdot \underline{Z}_3 = \underline{Z}_2 \cdot \underline{Z}_4$$

The amounts are subject to

$$Z_1 \cdot Z_3 = Z_2 \cdot Z_4$$

and for the phase angles

$$\varphi_1 + \varphi_3 = \varphi_2 + \varphi_4$$

The bridge is balanced when the product of the opposite complex resistances is equal.

Fig. 2.41 Capacitor in series
and parallel replacement circuit

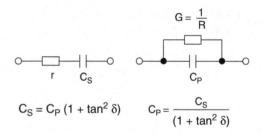

$$C_S = C_P (1 + \tan^2 \delta) \qquad C_P = \frac{C_S}{(1 + \tan^2 \delta)}$$

The losses of a capacitor are recorded by the measuring instrument either as series resistance $_{RS}$ or as parallel conductance G, regardless of how they occur. This is only a question of the measuring circuit. If the real loss components are to be determined, the measuring frequency must be selected accordingly: At very low frequencies the loss factor is determined by parallel losses alone, at very high frequencies only by series resistances. When the dissipation factor reaches $100 \cdot 10^{-3}$, the C-values measured in series and parallel equivalent circuits are no longer the same. Figure 2.41 shows both equivalent circuits and their conversion formulas. With a dissipation factor of $100 \cdot 10^{-3,}$ the series equivalent circuit C_S and parallel equivalent circuit C_P already differ by 1%. Since electrolytic capacitors have dissipation factors of up to 1.0 and higher, the equivalent circuit must also be defined to specify the nominal value. For example, a device that measures in the parallel equivalent circuit only provides half the value of the series capacitance at $\tan \delta = 1.0$, so that false conclusions may be drawn about compliance with the tolerance if this measurement condition is disregarded.

Figure 2.42 shows the circuit of a simple capacitance measuring bridge. The circuit is used for small (100 pF to 10 nF) and medium (1 nF to 1 µF) capacitances with negligible losses. It is also suitable for capacitive level measurement. The following applies

$$\frac{C_x}{C_4} = \frac{R_3}{R_2}$$

If the bridge is calibrated, the adjustable capacitor must have a value of $C_4 = 1$ nF. The capacitor C_4 has a value of 500 pF because it is set to 50%. This value with 50% is shown in the simulation.

Since in reality, no capacitor is loss-free, simply adjusting C_N is not sufficient to reach zero with the bridge output voltage ΔU, but a residual voltage remains. This makes a clear measurement of C_x difficult or even impossible. For this reason, each capacitance measuring bridge must be equipped with additional loss adjustment means to ensure that the voltages U_x and U_N are in phase. The adjustment resistors for the loss component are either connected in series with C_N or in parallel. The measurement result is then shown in either series or parallel connection. Only when the bridge voltage becomes zero can the full accuracy of the adjusted (calibrated) bridge elements be utilized. This also includes a sensitive zeroing instrument that displays only a few microvolts. Only tunable instruments with high amplification are suitable for this purpose, i.e., broadband voltmeters would also

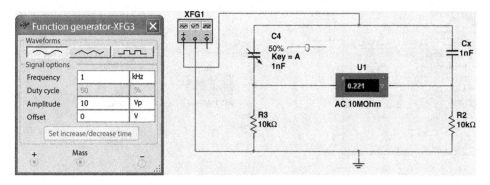

Fig. 2.42 Simple capacitance measuring bridge

display the unavoidable harmonics of the measuring generator, which pass through the bridge circuit undamped due to the frequency dependence of the DUTs. They influence the true minimum and thus an accurate measurement is impossible.

With bridge circuits in which resistors act as range-determining elements, it is possible to work at best up to several hundred kilohertz, because unavoidable stray components falsify the result. The measurement circuits used in practice differ from those shown in Fig. 2.42 in that the costly decade capacitors are replaced by a single fixed capacitor and the adjustment is carried out using less expensive decade resistors. It is remarkable that the adjustment element for capacitor losses directly provides the loss factor, related to a frequency, usually 1 kHz. This quality statement is more practical for the user than a resistance or conductance value.

The adjustment of a capacitance measuring bridge is done by alternately adjusting the potentiometers for the C-value and the loss adjustment until the bridge voltage becomes zero. This process is very fast if the bridge circuit is as shown in Fig. 2.42.

For the practical circuit with series or parallel set circuit, the adjustment can be tedious if tan δ comes into the range of $100 \cdot 10^{-3}$ and higher. This is especially the case when measuring electrolytic capacitors. The technician finds a tedious measurement where the minimum appears inaccurate and undefined in its position. Only when he has come across the true minimum after a systematic search, is there an exact calibration for an accurate measurement. In an executed device this disadvantage is eliminated by an automatic adjustment, which takes over the dissipation factor adjustment so that the technician only has to operate the C-adjustment and thus finds the correct minimum immediately without much effort.

2.4.5 Vienna and Vienna-Robinson Bridge

The Vienna Bridge is shown in Fig. 2.43 and is used for capacitance measurement by comparison with a precision capacitor C_4 with parallel resistor R_4.

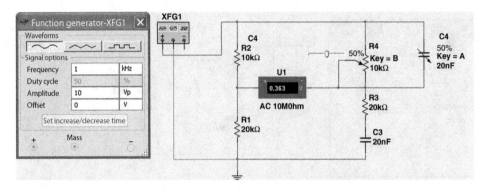

Fig. 2.43 Circuit for a Wien bridge

The Wien-Brücke is suitable for capacity measurements from 1 pF to 1000 μF and the error limit is 0.1%. With this measuring bridge also the loss factor *tan δ* can be measured and the error limit is only 1%.

In principle, all bridge circuits with frequency-dependent adjustment can be used for frequency measurement. The Wien-Bridge is calculated according to

$$\frac{C_4}{C_3} = \frac{R_1}{R_2} - \frac{R_3}{R_4} \quad \text{and} \quad \omega^2 = \frac{1}{R_2 \cdot R_3 \cdot C_3 \cdot C_4}$$

The Wien-Robinson-Bridge is constructed exactly like the Wien-Bridge and is also suitable for frequency measurement. This bridge was modified by Robinson by the additional conditions $R_1 = 2 \cdot R_2$, $C_3 = C_4 = C$, and $R_3 = R_4 = R$ and can be found in this form as the Wien-Robinson Bridge in practice. The adjustment conditions are

$$\omega = \frac{1}{R \cdot C} \quad \text{or} \quad f = \frac{1}{2 \cdot \pi \cdot R \cdot C}$$

Bridges of this type are designed with measuring ranges from 30 Hz to 100 kHz. They can also be used for distortion factor measurement. For the distortion factor measurement, the bridge is adjusted to the fundamental wave and the magnitude of the diagonal voltage formed from the harmonics is measured. If an RMS value former is used for this purpose and the measured value is related to the RMS value of the bridge input voltage, then the distortion factor can be specified directly as a percentage.

2.4.6   Maxwell Bridge

The Maxwell bridge is used to measure coils and capacitors. The induction measurements range from 1 μH to 10 H and the capacitance measurements range from 10 pF to 10,000 μF. Figure 2.44 shows the circuit for coils and capacitors, in which case only inductance is measured.

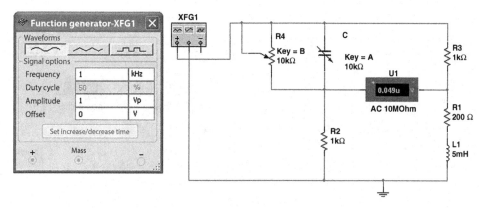

Fig. 2.44 Maxwell bridge for coils

Inductance measurements are usually carried out with the aid of the Maxwell-Vienna bridge. Modifications used in measurement technology are shown in Fig. 2.44. The adjustment conditions are

$$R_1 = \frac{R_2 \cdot R_3}{R_4} \quad \text{and} \quad L = R_2 \cdot R_3 \cdot C$$

It is advisable to use R_4 and C as adjustment elements, as they are independent of each other in both conditions, thus promoting convergence and speed of adjustment.

The following calculation results for Fig. 2.44

$$R_1 = \frac{R_2 \cdot R_3}{R_4} = \frac{1k\Omega \cdot 1k\Omega}{5k\Omega} = 200\Omega$$

$$L = R_2 \cdot R_3 \cdot C = 1k\Omega \cdot 1k\Omega \cdot 5nF = 5mH$$

Calculation and simulation are identical.

If the Maxwell bridge is modified, the measuring bridge in Anderson is reached, as shown in Fig. 2.45.

For the modified Maxwell bridge according to Anderson, the adjustment conditions are

$$R_1 = \frac{R_2 \cdot R_3}{R_4} \quad \text{and} \quad L = C \cdot \left[R_5 \cdot (R_1 + R_2) + R_1 \cdot R_4 \right]$$

If the resistor R_5 is used for the adjustment, L/C must always be greater than $R_1 \cdot R_4$, because the expression $R_5 (R_1 + R_2)$ cannot assume negative amounts.

The following calculation results for Fig. 2.45

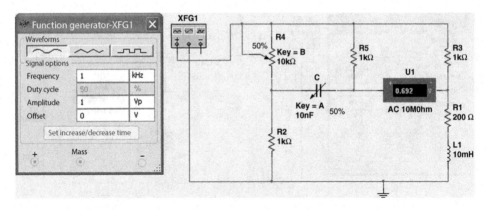

Fig. 2.45 Modified Maxwell Bridge to Anderson

$$R_1 = \frac{R_2 \cdot R_3}{R_4} = \frac{1\text{k}\Omega \cdot 1\text{k}\Omega}{5\text{k}\Omega} = 200\Omega$$

$$L = C \cdot \left[R_5 \cdot (R_1 + R_2) + R_1 \cdot R_4 \right] = 5\text{nF} \cdot \left[1\text{k}\Omega \cdot (200\Omega + 1\text{k}\Omega) + 200\Omega \cdot 1\text{k}\Omega \right] = 11\text{mH}$$

Calculation and simulation are identical.

2.4.7 Schering Bridge

The Schering bridge is used to determine the loss angle of capacitors. Figure 2.46 shows the circuit.

The adjustment conditions are as follows

$$R_x = R_1 \cdot \frac{C_4}{C_N}, \quad C_x = C_N \cdot \frac{R_4}{R_1} \quad \text{and} \quad \tan \delta_x = \omega \cdot R_4 \cdot C_4$$

The following calculation results for Fig. 2.46

$$R_x = R_1 \cdot \frac{C_4}{C_N} = 10\text{k}\Omega \cdot \frac{5\text{nF}}{10\text{nF}} = 5\text{k}\Omega$$

$$C_x = C_N \cdot \frac{R_4}{R_1} = 10\text{nF} \cdot \frac{1\text{k}\Omega}{10\text{k}\Omega} = 1\text{nF}$$

$$\tan \delta_x = \omega \cdot R_4 \cdot C_4 = 2 \cdot 3.14 \cdot 1\text{kHz} \cdot 1\text{k}\Omega \cdot 5\text{nF} = 31.4 \cdot 10^{-3} \Rightarrow 88°$$

Calculation and simulation are identical.

As is well known, the loss factor *tan δ* is the ratio of active to the reactive component of the apparent resistance or conductance. Except for electrolytic capacitors, the losses occur mainly in the dielectric. In the equivalent circuit diagram, they are represented as a

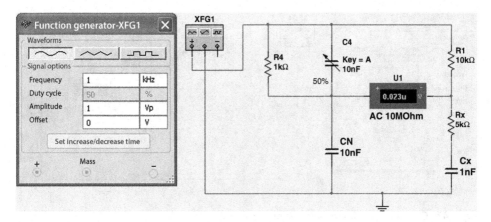

Fig. 2.46 Construction of the Schering Bridge

resistance R parallel to the loss-free imaginary capacity C. In addition to the resistor R, the insulation resistance R_{isol} must also be taken into account in parallel, which can only influence the loss factor at very low frequencies.

Due to the finite conductivity of the capacitor coatings and, above all, due to inadequate contact between the coatings and the connecting wires, further losses occur, which are represented in the equivalent figure by a series resistance r. However, their influence should be small compared to the dielectric losses, because they increase proportionally with frequency. An exception to this rule is electrolytic capacitors, whose losses are mainly caused by the resistance of the electrolyte, which also acts as a series resistor.

The dissipation factor of capacitors for oscillating circuits and metrological purposes should not be higher than about 10^{-3}. In filter circuits, where the impedance for the frequency to be short-circuited must be as low as possible, the losses of the dielectric play a minor role. Therefore, a loss factor of some 10^{-2} is permissible in these applications. Far more important, however, is a low series resistance, if possible in the milliohm range. An exception to this is again the electrolytic capacitors, whose loss factor can rise to 1.0 and above. However, such high values can only be expected for high-capacitance types, and *tan δ* averages 10^{-3}.

With AC capacitors that are loaded with strong currents, a too high dissipation factor due to the absorbed active power can lead to excessive heating and thus to a shortened service life.

2.4.8 Maxwell-Vienna Bridge

The Maxwell-Vienna bridge is used to determine small and medium inductances and the adjustment is frequency independent. Figure 2.47 shows the circuit.

The adjustment conditions are as follows

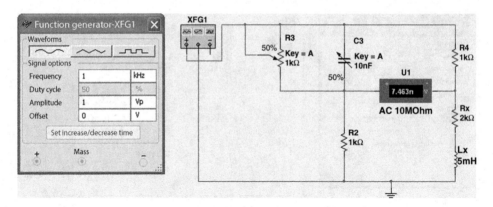

Fig. 2.47 Circuit of the Maxwell-Vienna bridge

$$R_x = \frac{R_2 \cdot R_4}{R_3} \quad \text{and} \quad L_x = R_2 \cdot R_4 \cdot C_3$$

The following calculation results for Fig. 2.47

$$R_x = \frac{R_2 \cdot R_4}{R_3} = \frac{1\text{k}\Omega \cdot 1\text{k}\Omega}{500\Omega} = 2\text{k}\Omega$$

$$L_x = R_2 \cdot R_4 \cdot C_3 = 1\text{k}\Omega \cdot 1\text{k}\Omega \cdot 5\text{nF} = 5\text{mH}$$

Calculation and simulation are identical.

Even with coil measurements, a bridge circuit always provides more accurate results. Because of their low natural resonant frequency and the frequency dependence of the losses, coils are unsuitable as bridge elements. They can, however, be replaced in a circuit specified by Maxwell by a capacitor diagonally opposite the DUT. Two examples of Maxwell bridge circuits are shown in Fig. 2.48. The equations are

$$\text{for high Q-values:} \quad \text{for low Q-values:}$$

$$L_x = R_2 \cdot R_4 \cdot C_3 \qquad L_x = R_2 \cdot R_4 \cdot C_3$$

$$Q_x = \frac{1}{\omega \cdot R_3 \cdot C_3} \qquad Q_x = \omega \cdot R_3 \cdot C_3$$

As the equations show, both capacitor C and one of the two diagonally opposite resistors can be used to adjust the inductance value; the latter is often preferred because it is less complicated. In Fig. 2.48 the resistor for the *L-adjustment* is marked R_2. The variable resistor R_3 compensates for the loss component. Its scale may be recorded in quality factor values because the adjusted resistance value is once directly proportional to Q and once inversely proportional to Q. By modifying the circuits, the losses could also be recorded as resistance values, but the quality factor is more meaningful and is preferred in practice.

Fig. 2.48 Maxwell bridge with quality indicator. *Left* for high Q values, *right* for low Q values

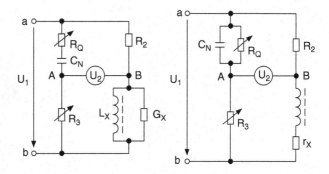

Losses in coils occur both in the winding and in the core and are represented in the equivalent picture by resistors in series to the loss-free imaginary inductance L. Individual components, such as the eddy current loss, are sometimes also replaced by a parallel resistor, but in the following only series, resistors will be considered. The following loss components are to be expected:

R_{Cu-}	Winding resistance at low frequencies
$R_{Cu\sim}$	Increase in winding resistance due to eddy currents
R_C	Loss resistance due to dielectric losses in the coil capacitance Cw
R_h	Core loss resistance due to remagnetization (hysteresis)
R_w	Core loss resistance due to eddy currents
R_r	Core loss resistance due to residual or after-effect losses

Each of these losses is responsible for a share of the total loss angle

$$\delta_{Total} = \delta_{Cu-} + \delta_{Cu\sim} + \delta_C + \delta_h + \delta_w + \delta_r$$

The quality factor of the coil gives the reciprocal of this total loss factor

$$Q = \frac{1}{\tan \delta_{Total}}$$

2.4.9 Frequency-Independent Maxwell Bridge

For some applications, the frequency-independent Maxwell bridge is also used and the circuit of Fig. 2.49 shows the setup.

The adjustment conditions are as follows

$$R_x = \frac{R_1 \cdot R_3}{R_4} \quad \text{and} \quad L_x = \frac{R_3 \cdot L_1}{R_4}$$

Please note that the potentiometer is divided into resistors R_3 and R_4. R_3 is on the right and R_4 on the left.

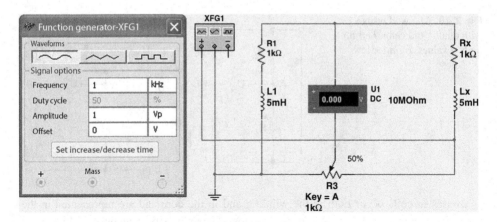

Fig. 2.49 Structure of the frequency-independent Maxwell bridge

The following calculation results for Fig. 2.49

$$R_x = \frac{R_1 \cdot R_3}{R_4} = \frac{1\text{k}\Omega \cdot 500\Omega}{500\Omega} = 1\text{k}\Omega$$

$$L_x = \frac{R_3 \cdot L_1}{R_4} = \frac{500\Omega \cdot 5\text{mH}}{500\Omega} = 5\text{mH}$$

Calculation and simulation are identical.

2.5　　Analog Switch

Wherever possible, the entire electronic system uses electronic analog switches in monolithic semiconductor technology rather than mechanical switches. Nevertheless, relays are still found at the outputs of a control circuit when high voltages and large currents need to be switched safely.

The main difference between relays and analog switches is the isolation between the signal control (relay coil to gate connection) and the signal to be controlled (contact to channel resistance). With semiconductor switches, the maximum analog signal depends on the characteristics of the FET or MOSFET transistors and the operating voltage. If an analog switch is used with an N-channel J-FET and there is no gate control, the switch is open. This also applies if the gate is driven with a negative voltage. The voltage between gate and drain or source is the "pinch-off" voltage. This behavior also applies to MOSFET technology. The analog signal is driven by the gate and thus a channel is built up (switch closed) or the channel is pinched off (switch open).

The contact resistances of relays are much lower than those of typical analog switches. However, the contact resistances do not play a significant role with the high input impedances of operational amplifiers, as the ratio is very large. In many circuits with analog

Fig. 2.50 Structure of an analog switch with the 4066

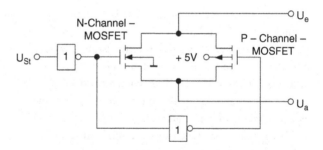

switches, contact resistances from 10 Ω to 1 kΩ do not cause serious errors in an electronic circuit, since these values are small compared to the high input impedances of operational amplifiers.

With the introduction of the 4066 devices from the CMOS standard series, the user had a bilateral switch available for switching analog and digital signals up to ±10 V with an operating voltage of ±12 V. Figure 2.50 shows the structure of an analog switch in CMOS technology.

In practice, the analog switch in Fig. 2.50 is referred to as a bilateral switch, since only simple protective measures are available internally. The technology used in this analog switch has not changed since 1970: Each channel consists of an N-channel and a P-channel MOSFET arranged in parallel on a silicon substrate and driven by the gate driver voltage of opposite polarity. The circuit of the CMOS device 4066 provides a symmetrical signal path through the two parallel resistors of source and drain. The polarity of each switching element ensures that at least one of the two MOSFETs will conduct at any voltage within the operating voltage range. Thus, the switch can process any positive or negative signal amplitude that is within the operating voltage range.

At high frequencies at the input, charge over coupling occurs from the control input via the gate channel or the gate-drain and gate-source capacitance to the input and/or output of this switch. Overcoupling is unpleasant in many applications, e.g. when a capacitor needs to be charged or discharged in a sample and hold or track and hold application. This behavior leads to disturbing offset voltages. For the 4066, the over coupled voltage is in the range of 30–50 pC, corresponding to 30–50 mV on a capacitor from 1 nF. This offset can be compensated by a signal of the same magnitude but reversed polarity, but this circuit is quite complex.

If a signal voltage may be present in an application without the 4066 operating voltage being properly present, the two internal MOSFET transistors will be destroyed.

Most analog switches used today operate according to the principle shown in Fig. 2.50. A CMOS driver drives the two MOSFET transistors, although an additional CMOS gate is required for the P-channel type. Both MOSFET transistors in the CMOS device 4066 switches simultaneously, whereby the parallel connection ensures a relatively even switch-on or contact resistance for the desired input range. The resulting resistance between U_e and U_a ranges from 100 Ω when switched on to 10 MΩ when switched off. The resistance between the gate terminal and the channel resistance reaches values up to 10^{12} Ω.

Due to the channel resistance, it is common practice to operate an analog switch in conjunction with a relatively high-impedance load resistor. In practice, an impedance converter is connected downstream of the analog switch. The load resistor can be very high-impedance compared to the switch-on resistor and other series resistors to achieve a high transmission accuracy. The transfer error is the input and output error of the analog switch connected to the load and the internal resistance of the voltage source. The error is defined as a percentage of the input voltage.

Using analog switches in data acquisition requires transmission errors from 0.1 to 0.01% or less. This can be achieved relatively easily by using buffer amplifiers with input impedances from 10^{12} Ω. Some circuits that are operated directly on an analog switch are already equipped with buffer amplifiers.

Cross talk is important in measurement technology. This is the ratio of output to the input voltage, whereby all analog channels must be parallel and switched off. The value of cross talk is usually expressed as output to input attenuation in dB.

Several leakage currents and capacitances must be taken into account for the operating status of analog switches. These parameters can be found in the data sheets and must be taken into account when operating analog switches. The leakage currents at room temperature are in the pA range and cause various problems only at higher temperatures. The internal capacitances also influence the crosstalk and the settling time of the analog switch.

In practice, error-protected analog switches must be used in the system electronics. If, for example, the internal or external power supply fails, the two MOSFET transistors in the analog switch will switch through. An output current flows through the two transistors, and the two transistors can be destroyed.

Overvoltages at the inputs of the analog switches cause an effect similar to the breakdown of the power supply. The overvoltage can turn a disabled analog switch to a state by setting the source terminal of the internal MOSFET to a higher potential than the gate connected to the power supply. As a result, the overvoltage not only puts a load on the connected sensors at the input of the measurement data acquisition but also on the devices after the analog switch.

For this reason, system configurations in which sensors in control systems are not connected to the same power supply are particularly at risk from power failure or overvoltage. To protect such a configuration, the developers concentrated on the analog switch, i.e. the point where the input signals first come into contact with the control logic. The first error protection circuits still contained discrete resistor and diode networks.

2.5.1   Switch Functions of the Analog Switches

Like a mechanical switch, each analog switch can process digital and analog signals in two directions, since they do not have an operating direction as is the case with digital gates. Depending on the control logic, these switches are closed (normally closed = NC) or open (normally open = NO) in the rest state. In general, a distinction is also made between the number of switchable contacts (single pole = SP, double pole = DP) and the type of contact

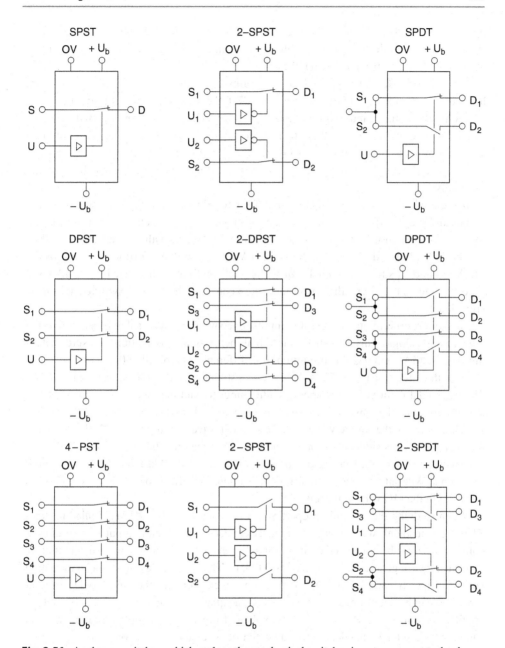

Fig. 2.51 Analogue switches, which replace the mechanical switches in measurement technology

(single throw ST) and switch (double-throw = DT). A change-over switch with one contact is therefore called "SPDT".

Figure 2.51 shows a comparison of the most important analog switches, distinguishing between the simple on-off switches, the multiple on-off switches, and the changeover switches. The voltages at the U inputs correspond to the TTL signal levels and these operate completely independently of the amplitudes of the analog signals. These analog signals must be within the operating voltage of the analog switch, but for reasons of linearity, it is advisable to try not to reach these limits.

The industry standard type DG201 contains e.g. four SPST switches, which are housed in one package together with a TTL/CMOS compatible control logic. Based on this component, the manufacturers developed the MAX331/2/3/4 analog switch series, which has operating currents that are 10 times lower (<10 µA) and offers all common switch combinations. A further advantage is that, unlike the original device, no separate logic operating voltage is required because the device generates this voltage internally.

If the analog switch is switched on, off, or toggled via the logic inputs, dynamic values and toggle effects occur. The order of magnitude in which these processes take place is in the nanosecond to microsecond range. Typical values are between 50 ns for the MAX334 and 500 ns for the DG201A for the switch-off time or 70 ns for the MAX334 and 1 µs for the DG201A for the switch-on time. In the case of the MAX333 type changeover switch, the difference between switch-off and switch-on time is significant.

To avoid short circuits between the inputs, one switch must be safely open before the other closes. A logic must provide for the "break-before-make" operation to ensure switching. For the MAX333 e.g. this minimum time difference is typically 50 ns.

With the circuit of Fig. 2.52, a dynamic investigation of the 4066 can be carried out. The sinusoidal input voltage is applied to the input S1 and the output D_1 is connected to the oscilloscope. The switchover between forward and reverse direction of the analog switch is done by the square wave generator which drives the input IN1. With a 1 signal, the analog switch is switched through, and with a 0 signal, it is blocked.

Important for this measurement is always a resistance of 10 kΩ to ground at the input and output. Without this external resistor, undesirable side effects occur, as various charges can form in the MOSFET channels.

When studying the dynamic properties, a frequency from 10 MHz is applied to input S1 and a sinusoidal AC voltage is switched between input S1 and output D_1. All unused connections must be connected to the ground. For the next measurement, connect input S1 to a square wave generator that generates 10 MHz. The signal delay time can be measured between input S1 and output D_1. With another measurement, the signal delay time at power-on can be determined. For this purpose, a square-wave frequency from 10 MHz is applied to input IN1 and a sinusoidal AC voltage from 1 MHz is applied to input S_2. You can now look at the signal delay time at switch-on when the oscilloscope shows the input and output voltage.

If the input IN1 is controlled with a 1 signal, the switch is "on", i.e. the channel resistance is relatively low impedance. If the input IN1 is connected to a 0 signal, the switch is disabled and the channel resistance is relatively high-impedance. The transition or channel

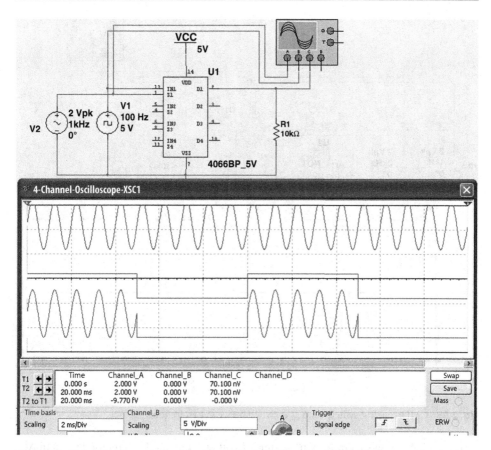

Fig. 2.52 Circuit for dynamic investigation of the CMOS analog switch 4066

resistance between input S1 and output D_1 is RK = 60 Ω, while the input resistance from IN1 to the channel is in the order of 10^{12} Ω.

The four separate analog switches in 4066 provide several switching options, as shown in Fig. 2.53. If a single-pole switch is implemented with the 4066, four separate switches with four control inputs can be constructed. The single-pole switches have two inputs and two outputs. Switching between the two outputs requires a NON-gate, which negates the control signal on input C accordingly. With a single 4066, two single-pole changeover switches can be implemented.

With the two-pole switch, two analog switches are connected in parallel and operated with a common control input. If the 4066 is used, two separate two-pole switches can be constructed. If a two-pole switch is required, two analog switches with a common input are used, and then two outputs are available. The NOT-gate provides the desired switching between the switches.

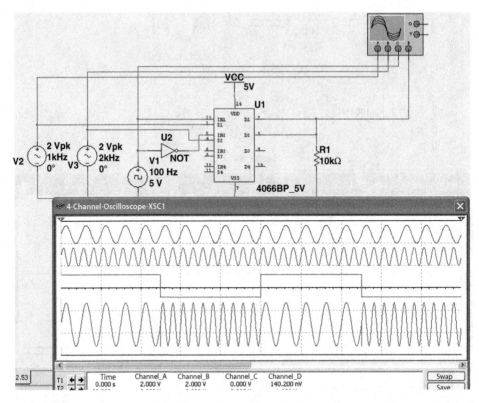

Fig. 2.53 CMOS analog switch 4066 as a switch between two frequencies

Figure 2.54 shows a circuit with different voltage sources. First on the left is an AC voltage source with 1 V_s and 1 kHz. Then follows a square wave voltage with 1 V, 1 kHz, and a duty cycle of 70%. The next voltage source AM generates an amplitude modulation and the voltage source FM a frequency modulation. These four voltage sources are connected to the four S inputs. Which voltage source is applied to the output of the analog switch depends on the switch position. The S point of the switch is connected to +5 V. The individual switch positions are connected to the IN inputs and a 1 signal means that the analog channels are switched through. This is prevented by the four resistors because if the switch does not provide a 1 signal, the IN inputs are at 0 signal and block.

2.5.2   Operational Amplifier with Digital Control

In analog circuit technology, digital control of operational amplifiers is often required to set the gain factor. In practice, two circuit variants can be used.

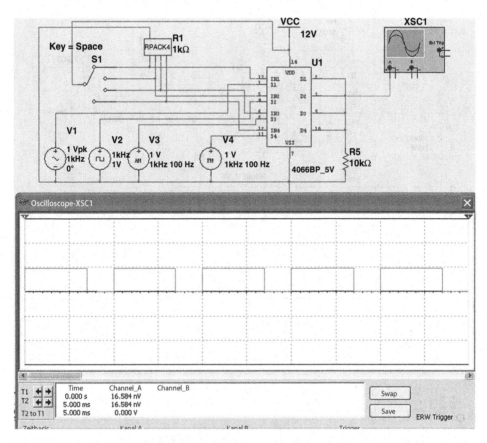

Fig. 2.54   Selection of voltage sources by an analog switch

In the circuit shown in Fig. 2.55, we have an operational amplifier with a constant resistance value R_5 in the feedback, while the input voltage is driven via a network of resistors with four different values of R_1, R_2, R_3, and R_4. The gain is calculated from

$$-U_a = U_e \cdot \frac{R_5}{R_1}$$

The input resistance can be controlled with the four switches. By controlling the analog switches, one can switch between the resistance values and select the appropriate gain. Important for this circuit are the high-impedance resistors at the inputs because you have to consider the channel resistance of the analog switches or you connect a compensation resistor from 100 Ω in series with the feedback resistor R_5. However, this compensation is not optimal, because due to the possibilities of connecting the four inputs in parallel, a control for the compensation resistor would also have to be done.

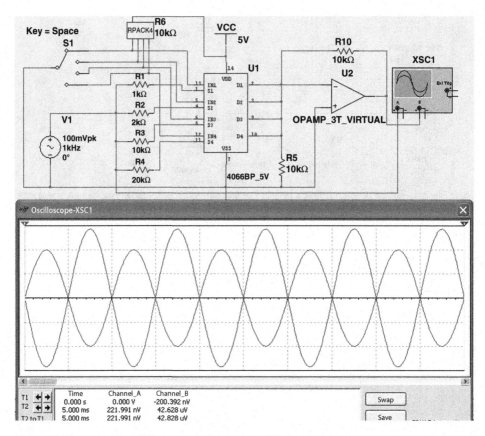

Fig. 2.55 Operational amplifier with digital control of the input resistors

In the circuit of Fig. 2.56, the digital control is located in the feedback of the operational amplifier and thus the gain can be controlled. Since the feedback resistors are implemented in dual code, a corresponding gain is obtained.

By controlling the analog switches, one can switch between the resistance values and select the appropriate gain. If the switch is set to position 1 and generates a 1-signal, the gain is $v = 1$, because both resistors are equal.

2.5.3   Sample & Hold Circuits

Due to the different and contradictory requirements, such as high speed and high accuracy, the sample and hold amplifier (Sample & Hold) is one of the most difficult analog circuits to control in practice. For example, the S&H unit is usually the biggest source of error in data acquisition systems. In addition to its use as a buffer before analog-to-digital converters, this circuit is mainly used for pulse-amplitude demodulation, automatic zero point

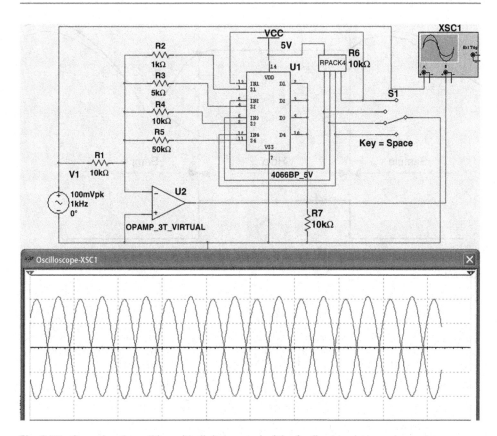

Fig. 2.56 Operational amplifier with digital control of the feedback resistors

correction in high-precision measuring systems, and for suppressing voltage peaks during switching operations.

In principle, an S&H unit has two tasks. In sampling mode (Sample), as shown in Fig. 2.57, the output voltage should follow the input voltage, comparable to the operation of a voltage follower. The distortion should be minimal in this mode (<0.01%), i.e., the difference between input and output voltage should be zero for each level and frequency.

In Hold mode, the current-voltage value should be stored for a certain time. A so-called hold capacitor $C_{S \, is}$ normally used as the storage element. Both operating modes are controlled by a digital command signal or by a time-synchronous clock pulse for pulse-amplitude demodulation.

The complete sample-and-hold circuit shown in Fig. 2.58 consists of the input amplifier with a high input resistance, which must supply the current for fast recharging of the holding capacitor $_{CH.}$ The analog switch couples the holding capacitor $_{CH}$ to the input amplifier in "Sample Mode" and in "Hold Mode" the analog switch disconnects the capacitor from the output of the input amplifier. The output amplifier has an extremely high input resistance and therefore only a very low discharge of the capacitor takes place.

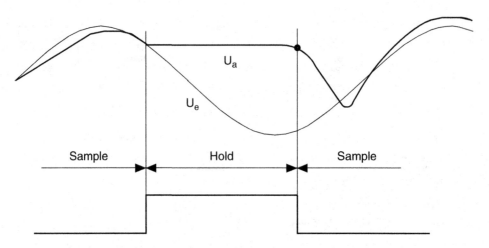

Fig. 2.57 Principle of a sample & hold circuit and during hold operation, the output voltage is kept constant

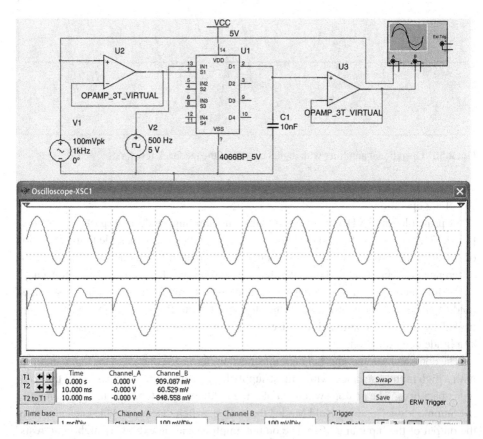

Fig. 2.58 Structure of a simulated sample & hold circuit with non-inverting input and output amplifiers

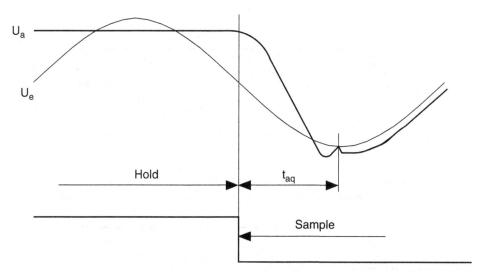

Fig. 2.59 Acquisition time for an S&H unit

The capacitance value of the holding capacitor directly influences the performance of the S&H unit. A hold capacitor with a low capacitance improves speed but also causes a loss of accuracy (Hold Step, Drop Rate). A capacitor C_H with a large capacitance worsens dynamic parameters such as bandwidth, slew rate, and acquisition time, and leads to high power consumption in AC signals due to recharging currents of the capacitor, which can cause a thermal fault.

The choice of the holding capacitor is problematic for these reasons. Hybrid sample and hold amplifiers already contain this capacitor. This eliminates any consideration. In general, Mylar or ceramic capacitors should not be used for monolithic circuits, as these have high dielectric and dynamic losses. Polystyrene (up to +70 °C), polypropylene (up to +85 °C) and polycarbonate or Teflon capacitors (up to +125 °C) are recommended.

To minimize leakage currents across the surface and internal resistance of the printed circuit in which the component is soldered, the connection of a hold capacitor should be shielded with a guard ring at output potential.

Since S&H devices are supposed to work with the accuracy of precision amplifiers and the speed of high-frequency amplifiers, an exact specification is important, especially for dynamic parameters. For this purpose manufacturers provide the following definitions:

- "Acquisition Time": The time t_{aq} required for the output voltage to settle after the command signal changes from "Hold" to "Sample". This process is shown in Fig. 2.59. This time depends primarily on the ability of the device to supply current to the hold capacitor. It also includes the value of the external hold element (recharging), the defined error limit (0.1% or 0.01%), and the amplitude of a required output voltage jump (usually 10 V).
- "Drop Rate": The drift rate of Fig. 2.60 is the rate of change of the output voltage U_a in holding mode. The ideal drift rate is 0 V/s, i.e., the stored voltage value remains

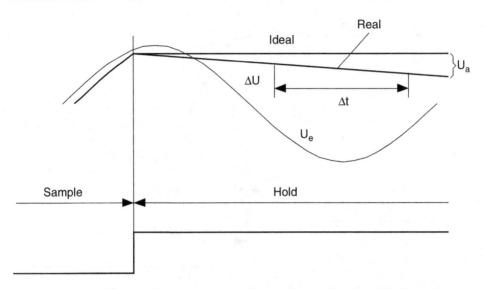

Fig. 2.60 Drift rate for an S&H unit

error-free for any length of time. However, the holding capacitor discharges due to leakage and base currents. The larger the leakage currents become, e.g. at high ambient temperatures, the greater the drift rate. A determining factor of this parameter is also the capacitance value of the holding capacitor since a larger capacitor takes longer to discharge for a given current. It is important here to consider the drift rate over the entire temperature range since in the case of an output amplifier constructed with a field-effect transistor the drift rate approximately doubles when the temperature rises to 10 °C.

The drift rate, another value for the holding capacitor C_H, which is specified in the datasheetdatasheet, is calculated as follows:

$$D_{R1} / D_{R2} = C_{H2} / C_{H1}$$

- "Drop Current" or drift current: With the specified drift current I_drop each user can calculate the "Drop Rate" D_R for a special holding capacitor $C_{H\,by}$ himself using the formula

$$D_R[\mathrm{V/s}] = \frac{I_\mathrm{drop}[\mathrm{pF}]}{C_\mathrm{H}[\mathrm{nF}]}$$

- "Aperture Time" or Opening Time: The time required to disconnect the hold capacitor from the input, open the analog switch, and thus save the output voltage when the control signal changes from "Sample" to "Hold" is called "Aperture Time" t_A, as shown in Fig. 2.61. The resulting settling time t_set only affects the speed, not the accuracy.
- "Aperture Jitter" or uncertainty of the aperture time: This specification is the deviation of the "Aperture Time" from mode change to mode change. To understand the

Fig. 2.61 Opening time t_A for an S&H unit

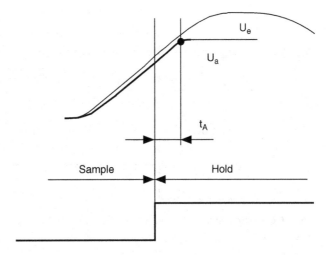

significance of this parameter, imagine that a 5 kHz signal changes by 0.05% in approximately 16 ns, which corresponds to 10-bit accuracy in an analog-to-digital converter.

- "Offset Voltage" or input error voltage: The offset voltage is the difference voltage between input and output in the "Sample" mode. It is comparable with the definition of operational amplifiers.
- "Hold Step": Voltage jump at the output when the control signal changes from "Sample" to "Hold". Figure 2.62 shows this process. The reason is a coupling of the control signal via stray capacitances and the internal capacitances of the analog switch to the hold capacitor. The "Hold Step" can be adjusted to zero in individual cases. However, in sampling mode, this parameter remains as an offset error.
- "Charge transfer" or transfer of charge: For different values of the holding capacitor, the "Charge Transfer" can be used to calculate the corresponding values for the offset error or "Hold Step" according to the following formula:

$$\text{Hold Step}\left[\text{mV}\right] = \frac{\text{Charge Transfer}\left[\text{pF}\right]}{\text{Haltekondensator}\left[\text{nF}\right]}$$

- "Zero Scale Error". This value can be measured by connecting the input of the module to 0 V and then switching the control line to "Hold". The voltage difference at the output against 0 V is called the "Zero Scale Error". This value includes the offset voltage and the "Hold Step".
- "Voltage Gain": The voltage gain is the ratio of input and output voltage in sampling mode, i.e. when driven over the input voltage range. This ratio is just under 1 for the configuration of the voltage follower. As can be seen from the formula below, this is primarily caused by the limited open-loop gain of the "Sample and Hold" and caused by the common-mode error of the input amplifier (gain $v = 1$). The ideal is the condition

$$v = 1 + \frac{R_1}{R_2}$$

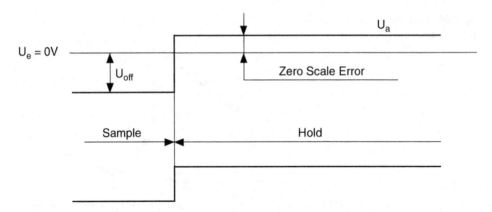

Fig. 2.62 Control signals for an S&H unit with "Hold Step", "Zero Scale Error" and offset voltage Uoff

But in reality

$$v = 1 + \frac{R_1}{R_2}\left(1 - \frac{1}{1+v_0}\right)$$

For example, an open-loop gain of $v_0 = 25$ V/mV results in a gain of $v = 0.99996$ or 0.4 mV deviation at the output with an input voltage of 10 V. Another error can be caused by the series connection of the output resistance of the "sample and hold" and the input resistance of the sequential circuit.

- "Gain Nonlinearity": This parameter describes not only the change of the gain due to the output level, the ambient temperature of the housing, the polarity of the input voltage, but also the most dominating common-mode error of the input stage in sampling mode.
- "Feedthrough" or crosstalk: This parameter specifies for hold mode how large an AC voltage applied to the input appears at the output. Internal stray capacitances are mainly responsible for this undesirable effect. The attenuation is inversely proportional to the value of the holding capacitor. Using a special circuit design and different adjustment procedures, the manufacturers of IC-S&H components try to solve the problems that occur. The offset voltage and the "Hold Step" are adjusted to a minimum value on the chip by "zener zapping" for a defined hold capacitor C_H. For other operating conditions and other values of the hold capacitor, the device can be adjusted with a "Zero" adjuster, i.e., at 0 V.

2.6   Analog-to-Digital and Digital-to-Analog Converters

When selecting AD and DA converters, several important considerations must be made before a circuit can be selected, purchased, and integrated into a system. To make the best possible selection, it is recommended that a list of the required features be drawn up in advance. This list should include the following key points:

- Converter type
- Resolution
- Transfer speed
- Temperature behavior

Since these considerations have already narrowed down the selection somewhat, several other parameters must be taken into account. These include the analog signal range at the input or output of the converter, the type of coding, the input and output impedance, the requirements for the operating voltage, the required digital interface to the microprocessor or microcontroller, the linearity error, the type of start and status signals for the AD converter, the influence of the operating voltages on the internal reference voltage sources, the dimensions of the housing and the weight. To make the selection process much easier, it is recommended to list all these parameters in order of importance. Last but not least, the price, delivery time, and good reputation of a manufacturer should not be forgotten.

2.6.1   The Structure of a Data Collection System

At the input of a data acquisition system is the sensor for converting the physical quantity into an electrical quantity. A larger output voltage is generated by a downstream analog measuring amplifier, which is then applied to an AD converter. The AD converter converts the analog voltage into a corresponding digital value which is then processed by a microprocessor or microcontroller. In Fig. 2.63 the two converters work as an interface to the outside world.

The AD-converter converts the current amplitude value of the input voltage into a digital format, after which it is processed in the microprocessor or microcontroller. Since a PC system in a process chain must not only record the state of a process but also control it, the calculated data must be output via a DA converter. This means that analog values are again available to the outside world for further processing, allowing different actuators to be controlled, e.g. in a closed control loop.

The task of a data acquisition system usually consists of quantization and digital processing as well as analysis and storage of the measurement data. In the field of practical measurement technology, there are usually 12-bit systems with eight input channels and a

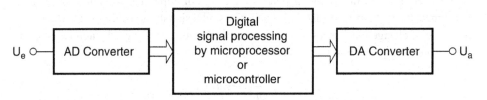

Fig. 2.63 Structure of a data acquisition system, consisting of an AD converter at the input, digital information processing, and a final DA converter

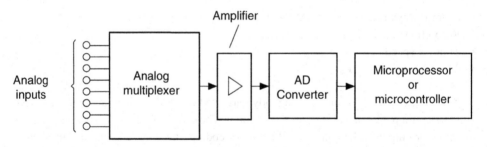

Fig. 2.64 Realization of a measurement data acquisition device without sample and hold unit

conversion speed adapted to the measurement task. Figure 2.64 shows a measurement data acquisition without a sample and hold unit. Measurement errors can occur if the conversion speed is too low.

An AD converter needs a certain small amount of time for the conversion process. This period required for conversion depends on several factors, such as the resolution of the converter, the conversion technique, and the speed of the components used in the converter. The conversion speed mainly depends on the time frame of the signal to be converted and the required accuracy.

The conversion time of the AD converter is often referred to as the aperture time. In general, the aperture time refers to the uncertainty span or the time window during a measurement. This results in an amplitude uncertainty and thus an error in the measurement if the signal amplitude changes during this time.

As Fig. 2.65 shows, the input signal of the AD converter changes by the amount ΔU during the aperture time t_a in which the conversion is performed. The error can be regarded as both amplitude and time error and both are linked together by the following relationship:

$$\Delta U = t_a \cdot \frac{dU(t)}{dt}$$

where $dU(t)/dt$ represents the time change of the input signal. It should be remembered that ΔU is the maximum error during the signal change. The actual error depends on how the conversion is performed in practice. At one point within the time frame $_{ta}$ the signal, amplitude corresponds exactly to the produced output codeword.

However, the error that occurs depends on the type of implementation procedure used. To estimate the real error, one assumes that a sinusoidal signal is to be digitized. The maximum rate of change reaches a sinusoidal signal at the zero-crossing. The amplitude error is calculated using the following equation

$$\Delta U = t_a \cdot \frac{d}{dt}\left(\hat{U} \cdot \sin \omega \cdot t\right)_{t=0}$$

Fig. 2.65 Aperture time and amplitude uncertainty for A/D converters and time t_a is the aperture time, ΔU shows the amplitude uncertainty

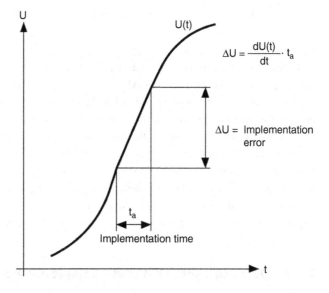

It follows from this

$$\Delta U = t_a \cdot \hat{U} \cdot 2 \cdot \pi \cdot f$$

If you define the factor ε as the ratio of the amplitude error to 2 times the peak value of the input voltage in LSB, you get

$$\varepsilon = \frac{\Delta U}{2 \cdot \hat{U}} = t_a \cdot \pi \cdot f$$

where $\varepsilon = 1/2^n$ can be represented for a 1-LSB error to

$$t_a = \frac{1}{\pi \cdot f \cdot 2^n}$$

This equation can be used to calculate the conversion time (acquisition time) for a defined amplitude error of 1 LSB (least significant bit or least significant bit). The conversion of this equation also allows the calculation of the maximum frequency of the input signal f_{max} for a given conversion time t_a and an amplitude error of 1 LSB:

$$f_{max} = \frac{1}{\pi \cdot t_a \cdot 2^n}$$

For a required amplitude error of 1/2 LSB, the permissible input frequency is halved.

Example: An AD converter with a resolution of 12 bit and maximum conversion time of 2 μs is to digitize a sine signal with 0 dB studio level (that is $4.36\ V_{ss}$). According to this equation, the permissible input frequency is

$$f_{max} = \frac{1}{\pi \cdot t_a \cdot 2^n} = \frac{1}{3.14 \cdot 2\,\mu s \cdot 2^{12}} = 38 Hz$$

With a required conversion accuracy of 1/2 LSB, f_{max} is halved to the value of 19 Hz. To digitize a sine wave voltage from 20 kHz, the opening time of the converter must not exceed 4 ns.

From this error discussion, it can be deduced that special converters with very small opening times must be used for AD conversion of signals with correspondingly high-frequency components. On the other hand, additional circuit measures can be taken to ensure that the instantaneous value of the analog voltage to be digitized is constant during the conversion. For this purpose, the instantaneous value is frozen to a certain extent when the sampling is triggered in a first step and then digitized in a second step. In most applications, also for cost reasons, an analog memory (sample and hold unit) is therefore connected upstream of the converter instead of an AD converter adapted to the signal speed. By connecting an analog memory upstream, a slower and therefore considerably less expensive AD converter can be used to solve the task at hand.

2.6.2 Data Acquisition Without a Sample and Hold Unit

A requirement often taken for granted is that the signal does not change by more than 1 LSB (least significant bit) within the maximum AD conversion time. In the case of a 12-bit converter that has a maximum vertical resolution of 4096 steps, the signal change during the conversion time must not exceed ±1 to 4096, which corresponds to $\pm0.024\%$. Otherwise, an AD converter with a lower resolution can be used. To ensure that the measuring signal does not change more than 1 LSB during the A/D conversion time, it must be frozen with the aid of a sample and hold unit. Exceptions are the direct conversion methods, such as the flash converter, which do not require a sample and hold unit. These devices are used for fast circuits (sampling frequencies >10 MHz with 8-bit resolution).

Using three measurement data acquisition systems that can be realized with different circuitry complexity, it will be shown where the limits of the different circuit principles are. Although the circuit shown in Fig. 2.64 is not common, it shows very clearly why a sample and hold unit is present in practically all AD systems. Example: A sinusoidal signal with ±10 V is applied to a converter system. The maximum applicable signal frequency is to be calculated, where the maximum error is 1 LSB. The AD converter used has a maximum converter speed of 1 MHz and a resolution in 12-bit format. To meet the general requirement for 1 LSB accuracy, the value of 1 LSB is first determined. The signal, which is $2 \cdot \hat{u}$ from peak to peak, is divided into $2^n - 1$ quantization stages, where n indicates the number of bits of the AD converter:

$$1\,\mathrm{LSB} = 2 \cdot \hat{u} / \left(2^n - 1\right) = 20\,\mathrm{V} / 4095 = 4.9\,\mathrm{mV}$$

To achieve an accuracy of 12 bits, the signal must not change by more than 4.9 mV within the maximum conversion time. To determine the maximum signal frequency of the present signal with the boundary conditions for the highest sampling rate and a 12-bit accuracy, the maximum voltage slope of the signal must be determined. The following generally applies to the maximum voltage slope of a sinusoidal signal

$$u(t) = \hat{u} \cdot \sin \omega t$$

$$\frac{du}{dt} = \hat{u} \cdot \omega \cdot \cos \omega t$$

$$\frac{du}{dt_{max}} = \left(\frac{du}{dt} \right)_{t=0} = \hat{u} \cdot \omega$$

The following approximation also applies

$$du = 1\,\text{LSB}$$

$$dt = 1\,\mu s \quad (\text{maximale Wandler-oder Umsetzzeit}).$$

This results in an input frequency for the maximum converter time t_u of

$$\frac{1\,\text{LSB}}{t_u} = \hat{u} \cdot \omega$$

$$\frac{4.9\,\text{mV}}{1\mu s} = 10\text{V} \cdot 2 \cdot 3.14 \cdot f$$

$$f = 1/3.14 \cdot 1\mu s \cdot 4095 = 77\text{Hz}$$

The value of the maximum converter time is very low with 77 Hz, whereby a very fast converter must be used. It can be seen that with direct conversion without sample and hold unit the maximum analog signal frequency must not be higher than 77 Hz, although a high sampling rate of 1 MHz is used. If the analog signal frequency is chosen higher, the error is increased accordingly, and the 12-bit resolution of the converter becomes a much lower measurement resolution.

If the input frequency is increased from 77 to 144 Hz, the voltage change Δu during the conversion time t_u is Δu of

$$\Delta u = \hat{u} \cdot \omega \cdot t_u$$

$$\Delta u = 9.67\,\text{mV}$$

, when the maximum input frequency is doubled, the error is also doubled, and the resolution of the system is based on an 11-bit format (2048 points). Since this is a linear relationship for sufficiently small changes Δu, the resolution is reduced by one-bit position each time, analogous to the multiplication of 77 Hz, which means a halving of the resolution or a doubling of the measurement error. As a result of this consideration, it is necessary to use sample and hold units in measurement data acquisition systems.

2.6.3 Time-Division Multiplexed Data Acquisition with Sample and Hold Unit

The task of a sample and hold unit is to freeze voltage values for a certain time. During this time, a downstream A/D converter can convert this frozen voltage into a digital value. Between the arrival of the hold command and its execution, a certain time t_{ap} (aperture time) passes, which, however, causes a temporal uncertainty t_{au} (aperture uncertainty). The "aperture time" is insignificant if it is the same for all input channels. Only the "aperture uncertainty" remains, which is usually in the range <200 ps.

Figure 2.66 shows a considerable improvement factor concerning the maximum signal frequency to be processed with

$$F = \frac{t_u}{t_{au}}$$

Example: A 12-bit AD converter with a conversion rate of 1 MHz is preceded by a sample and hold unit with ±50 ps "aperture uncertainty". This allows the highest signal frequency f_{max} to be processed with

$$f_{max} = \frac{t_u}{t_{au}} \cdot f = \frac{10^{-6}}{100 \cdot 10^{-12}} \cdot 77 \text{Hz} = 770 \text{kHz}.$$

This does not mean, however, that analog input signals can be processed up to 770 kHz since according to Shannon the frequency range of the measurement signal must remain limited to <500 kHz. The sample and hold unit significantly increases the permissible frequency range for the analog input signal. However, it also represents the part in a data acquisition system that contains the greatest sources of error. The demands on such a device are extreme since on the one hand, it has the speed of an RF amplifier and on the other hand, it should have the accuracy of a precision amplifier.

For the digitization of analog signals, sample and hold units are necessary and at the same time are decisive for the system accuracy. Practically all PC measuring cards on the market use a sample and hold unit. The initial general requirement of a measurement error from 1 to 4096 in a 12-bit system is thus, superficially speaking, fulfilled for at least one channel. However, the solution shown in Fig. 2.66, as it is used with simple PC measuring cards for cost reasons, has a number of decisive disadvantages. The problem lies in the analog multiplexed signal inputs, which perform a time-shifted sampling of the input channels. Since the measuring signal is amplified only after the analog multiplexers, all errors of the multiplexer are also amplified and are superimposed on the measuring signal. Besides, the measurement uncertainty is decisively dependent on the internal resistance of the signal source.

Example: An analog multiplexer has a temperature-dependent leakage current of I_{Don} = 300 nA per input, which flows into the signal source when the switch is closed and can cause considerable errors depending on the size of the source resistance. For example,

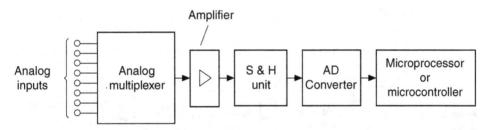

Fig. 2.66 Structure of a time-division multiplexed measurement data acquisition with several input channels, a common amplifier, a common sample and hold unit, and an A/D converter

the internal resistance of the signal source has a value of $R_i = 1$ kΩ. This results in a voltage U of

$$U = I_{Don} \cdot R_i = 300\,\text{nA} \cdot 1\,\text{k}\Omega = 0.3\,\text{mV}$$

which is a serious mistake.

If a high dynamic range is required or if sensors with mV signals are used, e.g. 16-bit AD converters are used. In this case, the amplifiers must be mounted before the analog multiplexer. Figure 2.67 shows the structure of a time-multiplexed data acquisition system with several input channels, each input having its preamplifier. If a sensor generates a voltage in the 100 mV range with an error of 0.3 mV, a measurement error of 0.3% results. However, the uncertainty due to the resolution of the AD converter is only 0.0015% in 16-bit format, i.e., an error of 0.3% of the sensor corresponds to about 200 LSB of the 16-bit AD converter.

It makes more sense here to connect a separate amplifier before the analog multiplexer and to provide an automatic offset adjustment, as shown in Fig. 2.67. Since the preamplifiers amplify the original signal to the voltage level of the system, the use of a 16-bit AD converter is reasonable, since the 12-bit range with a resolution of 4096 points is fully utilized. If, on the other hand, channel-specific preamplifiers are used, separate amplification per channel is also possible.

However, this arrangement only creates defined conditions for one channel. Several channels can be connected, but the recording caused by the multiplexer is not simultaneous. A temporal relationship can only be established by accepting considerable measurement errors. Even simple temporal cursor measurements are afflicted with considerable errors. The allocation of channels, as required for example for power determination, leads to large errors. If only sinusoidal quantities are used for consideration, the reactive power Q is given at a phase angle φ with $Q = U \cdot I \cdot \sin \varphi$. The time-multiplexed sampling of current and voltage acts like an additional phase shift of $\Delta\varphi$. For small angles φ, $\sin \varphi \approx \Delta\varphi$ applies approximately, so that the relative error F_r to $F_r{}^* = \Delta\varphi/\varphi$ can be determined.

Example: If a sinusoidal current and a sinusoidal voltage with a frequency of 50 Hz and a phase angle of 5° between current and voltage are detected with a system of Fig. 2.67 with a sampling frequency of 5 kHz, $\Delta\varphi$ is made up of

$$\Delta\varphi / 360° = 0.2\,\text{ms} / 20\,\text{ms}$$

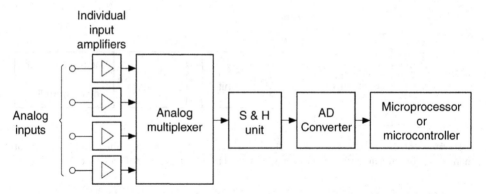

Fig. 2.67 Structure of a time-division multiplexed measurement data acquisition with several input channels, each input using a separate preamplifier

predictable. It applies

$$\Delta\varphi = 3.6°$$

This causes a relative error of $F_r^* = 72\%$. This shows that signals that are to be brought into a temporal relationship to each other or offset against each other require simultaneous sampling by channel-specific sample and hold units. This even applies to many applications where only very slow signals such as temperatures have to be acquired and related to each other.

2.6.4 Simultaneous Acquisition of Measurement Data with a Sample and Hold Unit

Only measurement data acquisition systems with separate sampling and holding units can bring several recorded signals into a time reference to each other or offset the signals against each other. Although both Fig. 2.67 and Fig. 2.68 use sample and hold units, the major difference in Fig. 2.67 is that the individual channels can be sampled simultaneously. As long as several channels are to be recorded and a time relationship between the channels never has to be established in the analysis, both methods are identical. At the moment when time-correlated measurements of several channels have to be performed, considerable errors occur when switching Fig. 2.67. Although the signals, considered in isolation, all have the same error, the serial sampling of the analog multiplexer causes them to be in phase with each other by at least one sampling interval. However, this phase shift can be very disadvantageous in a PC precision measuring card.

The circuit shown in Fig. 2.68 shows the capability for simultaneous sampling of the input channels. This is called the SS&H unit (simultaneous S&H). Here, the input signals are first brought to a certain voltage level with individual preamplifiers. Afterward, all input channels are sampled in parallel by individual sample and hold units and switched in

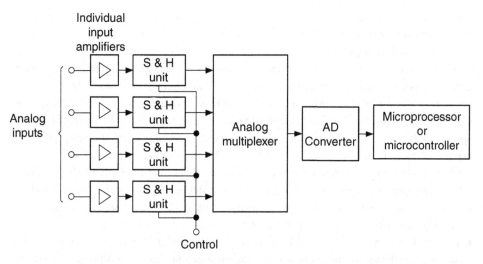

Fig. 2.68 Structure of a simultaneous measurement data acquisition with several input channels, each input having a separate preamplifier with a downstream sample and hold unit

analog form to the analog multiplexer, which then transfers them in time sequence to the A/D converter for conversion. The main advantage of this method is that the individual channels are sampled simultaneously. Only in this case a time reference, i.e. a cursor measurement between the channels or a calculation of the channels, is possible without problems.

2.6.5 Antialiazing Filter

For signal conditioning, the use of anti-aliasing filters is conceivable. Whether each low-pass filter should be used to avoid anti-aliasing effects depends on whether the measurement signal still has frequency components above half the sampling frequency or not. A frequently used solution for bypassing channel-specific anti-aliasing filters is to limit the bandwidth of the input circuit to typically 1/3 of the maximum sampling frequency. Although this solution is optimal for the first block, it must be considered that each analog filter attenuates not only the measurement noise but also the measurement signal. If the highest analog signal frequency in such a case is e.g. 1/3 the maximum sampling frequency, the measurement error, in this case, is −3 dB, i.e., −30%.

Such errors can never be tolerated in a high-resolution measuring system. On the other hand, it must be remembered that the cut-off frequency of such a band limitation cannot be changed and thus has no effect on slow sampling rates. From a metrological point of view, it is cleaner to use input circuits with a bandwidth that is a multiple of the maximum sampling rate to generate only slight signal distortions when using the maximum sampling rate. If low-pass filters (anti-aliasing filters) are needed in the input circuit, they should be programmable in order, slope limitation and characteristics. This ensures that only those high-frequency components (interference) are filtered that cannot physically originate from the measured object.

Filters that only serve to limit bandwidth are called anti-aliasing filters. Artificially generated noise is always periodic, whereby the mains hum voltage must be taken into account as an example, and this can only be suppressed with a bandstop. In contrast, general noise is randomly distributed in amplitude and frequency over the entire frequency spectrum. Noise sources are often sensors, resistors, etc. and even preamplifiers. This noise is reduced by reducing the bandwidth of the system to the necessary level.

As already noticed with the amplifiers, there are no perfect filters that solve all problems. Compromises must always be made when selecting the appropriate filter. Ideal filters, which are often based on theoretical analyses, have a horizontal line up to the cut-off frequency and then drop vertically with infinite attenuation. However, these are mathematical filters that cannot be realized in practice. The developer is usually given the cut-off frequency and the minimum attenuation. Damping and phase position depends on the filter characteristics and the number of poles. Known filter characteristics are e.g. those according to Butterworth, Chebyshev, and Bessel. Before making any decision on the filter type, the user must pay close attention to the overshoot characteristics and the steepness of the attenuation.

The basic circuit structure shown in Fig. 2.66 can only be used to a very limited extent for accurate and multi-channel measurement data acquisition. If the time reference between individual channels is not of interest, a circuit structure as shown in Fig. 2.67 can be used. If the time reference between individual channels is relevant for subsequent analyses and/or the input channels are to be offset against each other, the basic circuit structure as shown in Fig. 2.68 is used.

2.6.6 Systems for Signal Sampling

In practice, analog input signals are sampled periodically. The pulse diagram in Fig. 2.69 shows the signal sampling at individual points in time.

Diagram (a) shows the input voltage. The sequence of sampling pulses (b) is realized by a fast-acting switch, which turns on the analog signal for a very short period and stays off for the rest of the sampling period. The result of this fast sampling is identical to the multiplication of the analog signal by a pulse train of the same amplitude, resulting in a modulated pulse train (c). The amplitude of the original signal is contained in the envelope of the modulated pulse train.

If this sampling switch is now supplemented by a capacitor in a sample and hold unit, the amplitude of each sample can be stored for a short time. The result is a useful reconstruction (d) of the original analog input signal.

The purpose of sampling via an analog multiplexer is the efficient use of information processing and data transmission systems. A single data transmission link can be used to

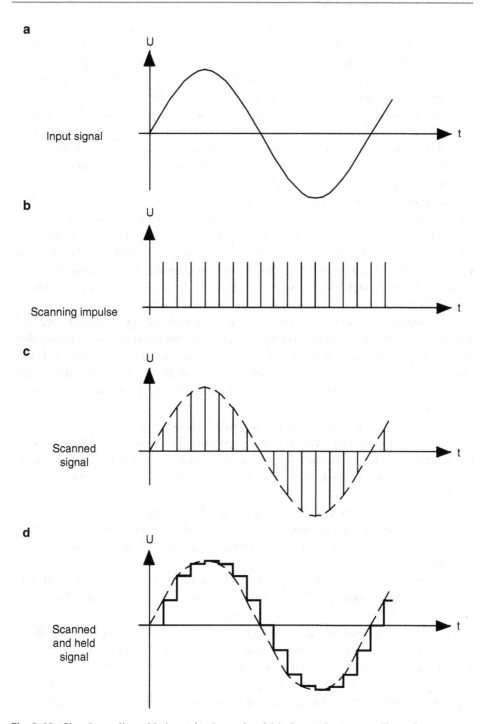

Fig. 2.69 Signal sampling with the analog input signal (**a**), the continuous sampling pulse sequence (**b**), the sampled signal (**c**), and the held signal (**d**)

transmit a whole series of analog channels. It would be very uneconomical to occupy a complete data transmission chain for the continuous transmission of a single signal. Similarly, a data acquisition and distribution system is used to measure and monitor the many parameters of a process control system. This is also done by scanning the individual parameters through the periodic acquisition of the control inputs.

In data conversion systems, it is common to use a single but expensive AD converter with high speed and high accuracy and to have it sample several analog input channels in multiplexed operation. An important and fundamental question when considering a sampling system is the following: How often must an analog signal be sampled in order not to lose any information during reconstruction?

All useful information can be detected from a slowly changing signal if the sampling rate is so high that there is no or almost no change in the signal between samples. It is equally obvious that if the signal changes rapidly between samples, important information may be lost. The answer to this question is given by the well-known sampling theorem, which is as follows: If a continuous signal of limited bandwidth does not contain any frequency components higher than f_C (corner frequency), the original signal can be restored without interference losses if the sampling is at least at a sampling rate of $2 \cdot f_C$.

The sampling theorem can be represented or explained with the frequency spectrum shown in Fig. 2.70. Diagram (a) shows the frequency spectrum of a continuous, bandwidth-limited analog signal with frequency components up to the corner frequency f_C. If this signal is sampled at the rate f_S, the modulation process shifts the original spectrum to the points f_S, $2 \cdot f_S$, $3 \cdot f_S$ etc. beyond the original spectrum. Part of this resulting spectrum is shown in the diagram (b).

If now the sampling frequency f_S is not selected high enough, a part of the spectrum belonging to f_S will overlap with the original spectrum. This undesirable effect is known as frequency folding. When the original signal is restored, the overlapping part of the spectrum causes indefinable interference in the new signal that cannot be eliminated even by filtering.

Figure 2.70 shows that the original signal can only be restored without interference if the sampling rate is set high enough that $f_S - f_C > f_C$. Only in this case, the two spectra are clearly adjacent to each other. This again proves the assertion of the sampling theorem, according to which $f_S > 2 \cdot f_C$.

Frequency overlap can be prevented in two ways: firstly, by using a sufficiently high sampling rate and secondly, by filtering the signal before sampling to limit its bandwidth to $f_S/2$.

In practice it can always be assumed that depending on high-frequency signal components, noise, and non-ideal filtering, a small frequency overlap will always occur. This effect must be reduced to a negligible amount for the specific application by setting the sampling rate high enough. The necessary sampling rate may be much higher in the real application than the minimum required by the sampling theorem.

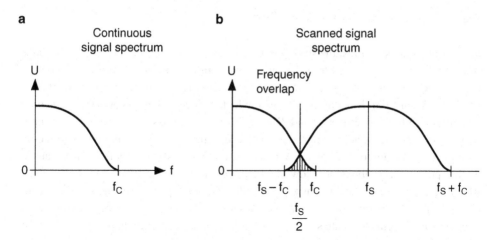

Fig. 2.70 Frequency spectrum of a continuous band-limited signal (**a**) and the frequency spectrum resulting from the sampling theorem (**b**)

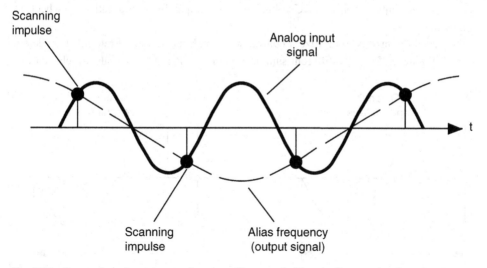

Fig. 2.71 Generation of an apparent frequency if an unsuitable sampling rate is selected

The effect of an unsuitable sampling rate on a sinusoidal input voltage is shown in Fig. 2.71. An apparent frequency (alias frequency) is the result of trying to restore the original signal. In this case, a sampling rate of slightly less than twice per waveform results in the low-frequency sinusoidal oscillation. This apparent frequency can be clearly distinguished from the original frequency. The figure also shows that a sampling rate of at least twice per curve, as required by the sampling theorem, makes it easy to restore the original curve.

2.6.7 Theorem for Signal Sampling

In moving pictures in films, the spoked wheels of moving carriages often seem to behave strangely. The wheels either turn too slowly, stop or turn backward. This effect is comparable to the creation of aliasing frequencies in electronic systems and can be explained similarly. When a film is being shot, for example, the video camera records the scene with 50 fields or 25 full frames per second. During this time, the passing car generates an input frequency that corresponds to the rotational speed of the wheel in revolutions per second, multiplied by the number of spokes. When the car changes speed, this frequency, caused by the spokes, varies above or below the integer multiples of the frame rate. Later, when watching the film, the wheel seems to stand still when the "spoke frequency" is an integer multiple of the frame rate. At spoke frequencies just below an integer multiple of the frame rate, the wheel seems to rotate slowly backward, whereas at slightly higher frequencies it rotates slowly forwards.

An analog-to-digital converter, for example, which samples a pure sine wave signal exactly once per period at constant intervals, will measure the same value at each conversion and thus output a constant DC voltage signal, comparable to the stationary wheel on a moving car.

So when the task is to digitize a continuous signal, the sampling rate must be determined. Figure 2.72 shows different sampling intervals. If these intervals are chosen too

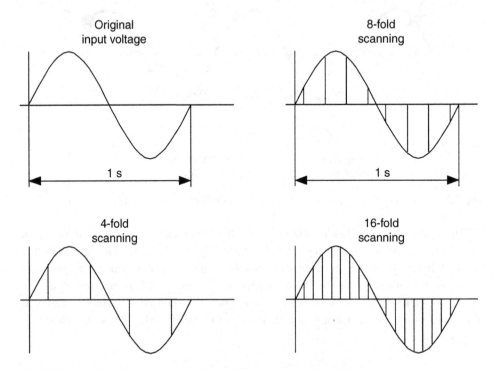

Fig. 2.72 Sampling a sinusoidal signal by a different number of samples

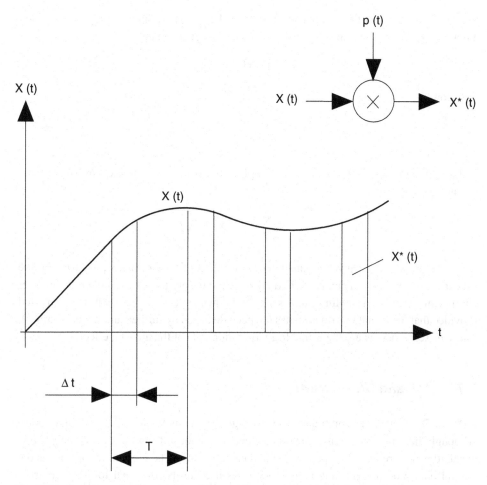

Fig. 2.73 Sampling of a dynamic signal by an AD converter with $x(t)$ as the original signal, $p(t)$ as a sampled signal, and $x\,{}^*(t)$ as a sampled signal

short, the individual samples show an unnecessarily high redundancy to each other and also make the entire measurement setup more expensive. However, if the intervals are too large, some signal information is lost and the original signal can only be reconstructed inaccurately.

From Fig. 2.73 it can be seen that the result of the sampling can be interpreted as a multiplication of the original signal by the sampling signal. From this point of view, a representation of the sum signal in the frequency response is of particular importance. Since the series of sampling pulses is a periodic function, it can be described as a Fourier series:

$$p(t) = \sum_{n=-\infty}^{n=+\infty} \vec{c}_n \cdot e^{j(n\omega_0 t)}$$

where $\omega_0 = 2 \cdot \pi/t$ is the angular frequency of the sampling signal. With the sampling angular frequency $_{\omega 0}$, the sum signal of the sampled signal can be

$$x^*(t) = x(t) \cdot p(t)$$

as follows:

$$x^*(t) = \sum_{n=-\infty}^{n=+\infty} \vec{c}_n \cdot x(t) \cdot e^{j(n\omega_0 t)}$$

Using the displacement theorem, the Laplace transform $x^*(s)$ of the sum signal $x^*(t)$ can now be formed:

$$x^*(s) = \sum_{n=-\infty}^{n=+\infty} \vec{c}_n \cdot x(s - in\omega_0)$$

The transform $x^*(s)$ gives an unaltered description of the original signal $x(t)$. For the special case n = 0, $x^*(s)$ can be re-normalized by multiplying it by c_0. Since the Fourier coefficient, c_0 of the sampling pulse series is known, $x^*(s)$ can be easily reconstructed provided that the signal information has not been distorted by further summands [$c_0 x^*(s)$]. These mixed products are the actual triggering moments of the aliasing effect.

2.7   AD- and DA-Converter

AD and DA converters communicate with digital systems using suitable digital codes. Although there is a wide range of possible codes to choose from, in practice only a few standard codes are used for operation with data converters. The most common code is the natural binary or straight binary code, which is used in its fractional form to represent a number

$$N = a_1 \cdot 2^{-1} + a_2 \cdot 2^{-2} + a_3 \cdot 2^{-3} + \ldots + a_i \cdot 2^{-n}$$

where the coefficients ai may be = 0 or 1 and n represents a natural number.

2.7.1   Natural Binary Code

A binary number is usually written as 0.110101. However, in the codes of the data converters, the decimal place is omitted and the code word is represented as 110101. This codeword represents a range of the converter's end range value.

The binary codeword 110101 of this example, therefore, represents the decimal numerical value of $(1 \cdot 0.5) + (1 \cdot 0.25) + (0 \cdot 0.25) + (0 \cdot 0.25) + (1 \cdot 0.0625) + (0 \cdot 0.03125) + (1 \cdot 0.01562) = 0.828125$ and corresponds to 82.8125% of the full-scale value at the

Table 2.2 Resolution, number of states, LSB weighting, and dynamic range for AD/DA converters

Bit resolution n	Number of states 2^n	LSB weighting 2^{-n}	Dynamic range in dB
0	1	1	0
1	2	0.5	6
2	4	0.25	12
3	8	0.125	18.1
4	16	0.062	24.1
5	32	0.03125	30.1
6	64	0.015625	36.1
7	128	0.0078125	42.1
8	256	0.00390625	48.2
9	512	0.001953125	54.2
10	1024	0.0009765625	60.2
11	2048	0.00048828125	66.2
12	4096	0.000244140625	72.2
13	8192	0.0001220703125	78.3
14	16.384	0.00006103515625	84.3
15	32.768	0.000030517578125	90.3
16	65.536	0.000012587890625	96.3
17	131.072	0.00000762939453125	102.3
18	262.144	0.000003814697265625	108.4
19	524.288	0.0000019073486328125	114.4
20	1.048.576	0.00000095367431640625	120.4

output of the converter. If a maximum voltage end value of +10 V is defined for this consideration, this code word represents an output voltage of +8.2812 V.

The natural binary code belongs to a class of codes known as positively weighted. Each coefficient has a special weighting here, negative values do not occur. The bit furthest to the left has the highest weighting, namely 0.5 of the full range value, and is known as the "most significant bit" or MSB. The bit furthest to the right, on the other hand, has the lowest weighting, namely 2^{-n} of the end range and is therefore called the "least significant bit" or LSB. The bits in a codeword are numbered from left to right with 1 to n. Table 2.2 shows the resolution, number of states, LSB weighting, and dynamic range for data converters.

The analog value of an LSB for the binary code is

$$\mathrm{LSB}\left(\mathrm{Analog\ \ value}\right) = \frac{\mathrm{FSR}}{2^n}$$

where FSR (full-scale range) is the measuring range between the minimum and maximum measuring voltage. The maximum voltage is called a "full scale" (FS). The following applies to a unipolar converter: FS = FSR.

The dynamic range in dB of a data converter is calculated from

$$DR[dB] = 20 \cdot \lg 2^n$$
$$= 20n \cdot \lg 2$$
$$= 20n \cdot (0.301)$$
$$= 6.02 \cdot n$$

where DR is the dynamic range, n is the number of bits and 2^n represents the number of output states of the converter. Since 6.02 dB corresponds to a factor of 2, the resolution of the converter, i.e. the number of bit positions, must simply be multiplied by 6.02 to determine DR. A 12-bit converter thus has a dynamic range of 72.2 dB.

2.7.2   Complementary Binary Code

An important point that must be observed is the following: The maximum value of a digital code consisting of loud 1 signals does not correspond to the analog maximum value. The maximum analog value is 1 LSB below the final value and is calculated from the relationship corresponding to FS (full scale or final value) multiplied by $1 - 2^{-n}$. Consequently, a 12-bit converter with an analog voltage range from 0 to +10 V has a maximum code of 1111 1111 1111 for a maximum analog value of $+10 \text{ V} \cdot (1 - 2^{-12}) =$ +9.99756 V. In other words, the maximum analog value of a converter, according to a code word made up of all 1 signals, can never quite reach the value defined as the actual final analog value.

Table 2.3 shows the natural and complementary binary code for a unipolar 8-bit converter with an analog voltage range from 0 to +10 V at its output. The maximum analog value for this converter is +9.961 V or $+10 \text{ V} - 1$ LSB. As shown in the table, the value for 1 LSB is 0.039 V.

2.7.3   Codes for AD and DA Converters

In addition to the natural binary code, several other codes are used in AD and DA devices, in particular the offset binary code, the 2's complement code, the binary coded decimal code (BCD), and their complementary versions. Each code has specific advantages in

Table 2.3 Direct and complementary binary coding of a unipolar 8-bit converter

Value of FS-LSB	+10 V FS	Binary code	Complementary binary code
+FS −1 LSB	+9961	1111 1111	0000 0000
+3/4 FS	+7500	1100 0000	0011 1111
+2 FS	+5000	1000 0000	0111 1111
+1/4 FS	+2500	0100 0000	1011 1111
+8 FS	+1250	0010 0000	1101 1111
+1 LSB	+0.039	0000 0001	1111 1110
0	0.000	0000 0000	1111 1111

Table 2.4 Bipolar coding for data converters

The parliamentary group of FS	±5-V-FS	Offset binary	Complementary offset binary	2's complement	Sign-Mag. binary
+FS −1 LSB	+4.9976	1111 1111	0000 0000	0111 1111	1111 1111
+3/4 FS	+3.7500	1110 0000	0001 1111	0110 0000	1110 0000
+1/2 FS	+2.5000	1100 0000	0011 1111	0100 0000	1100 0000
+1/4 FS	+1.2500	1010 0000	0101 1111	0010 0000	1010 0000
0	0.0000	1000 0000	0111 1111	0000 0000	1000 0000
−1/4 FS	−1.2500	0110 0000	1001 1111	1110 0000	0010 0000
−1/2 FS	−2.5000	0100 0000	1011 1111	1100 0000	0100 0000
−3/4 FS	−3.7500	0010 0000	1101 1111	1010 0000	0110 0000
−FS + 1 LSB	−4.9976	0000 0001	1111 1110	1000 0001	0111 1111
−FS	−5.0000	0000 0000	1111 1111	1000 0000	

certain applications. For example, the BCD code is used when an interface to a digital display has to be established, as is the case with digital seven-segment displays or multimeters. The 2's complement code is used in computers for arithmetic and logic operations, the offset binary code is used for bipolar analog measurements.

In connection with data converters not only the digital codes are standardized, but also their analog voltage ranges. Most converters use the unipolar voltage range from 0 to +5 V and from 0 to +10 V, a few also use the negative range from 0 to −5 V or from 0 to −10 V. The standard bipolar voltage ranges are ±2.5 V, ±5 V and ±10 V. Table 2.4 shows a comparison of the most important codes.

The 2's complement code has the property that the sums of its positive and negative codes for the same absolute value always result in louder 0 values plus a carry. This characteristic allows this code to be used when arithmetic calculations have to be performed with the measurement results. Note that the complementation of the MSB is the only difference between the 2's complement and offset binary codes. With bipolar coding, the MSB here becomes the sign bit.

The sign-magnitude binary code, on the other hand, is hardly ever found. This code has identical code words for identical positive or negative absolute values. They differ only in the sign bit. As Table 2.3 shows, this code also has two possible code words for the value zero:1000 0000 and 0000 0000. The two are usually distinguished as 0± and 0−. Because of this characteristic, the code has maximum analog values of +(FS − 1 LSB) and therefore does not reach either +FS or −FS.

2.7.4   BCD Coding

The BCD code, also known as 8-4-2-1 code, is a pure binary code shortened to one decade. It can be used for arithmetic and control tasks (counting) or when interfaces are implemented for digital display devices. Its significance corresponds to the first four digits of the

Table 2.5 BCD and complementary BCD code in 12-bit format, with three decimal places

The parliamentary group of FS	+10-V-FS	binary-coded decimal code	Complementary BCD code
+FS − 1 LSB	+9.99	1001 1001 1001	1001 0110 0110
+3/4 FS	+7.50	0111 0101 0000	1000 1010 1111
+1/2 FS	+5.00	0101 0000 0000	1010 1111 1111
+1/4 FS	+2.50	0010 0101 0000	1101 1010 1111
+1/8 FS	+1.25	0001 0010 0101	1110 1101 1010
+1 LSB	+0.01	0000 0000 0001	1111 1111 1110
0	0.0	0000 0000 0000	1111 1111 1111

pure binary code, the dual code. The BCD code of decimal numbers from 0 to 9 is formed by real tetrades and from A (10) to F (15) by the so-called pseudotetrades.

Table 2.5 shows the BCD and complementary BCD coding for a data converter with three decimal places (digits). These codes are used for integrating AD converters as used in digital panel displays, digital multimeters, and other applications with the decimal display. Here four digits are used to represent one decimal place. The BCD code is positively weighted. Its use is relatively inefficient since in each group of four bits only 10 of the 16 possible output states are used. The analog value of an LSB for the BCD code is

$$LSB(Analog \quad value) = \frac{FSR}{10^d}$$

where FSR represents the scale end range and d the number of decimal digits. So if, for example, three digits are to be represented and the voltage range is $U_e = 10$ V, the value of an LSB results from

$$LSB(Analog \quad value) = \frac{10V}{10^3} = 0.01V = 10mV$$

The BCD code is usually provided with an additional overflow bit, which has the weighting of the full-scale value. This increases the range of the AD converter by 100%. A converter with a decimal scale end range of 999 receives the new range of 1999 through the overflow bit, which means a doubling of the original range. In this case, the maximum output code is 1 1001 1001 1001. The additional range is usually referred to as 1/2 digit. So the resolution of the AD converter is 3 1/2 digit.

If the end range is stretched by a further 3 3% similarly, a new end range of 3999 is obtained, in this case, we speak of a converter with a resolution of 3 3/4 digit. By adding its two overflow bits, the output code for the scale end range appears as 11 1001 1001 1001. If the BCD code is to be used for a bipolar measurement, an additional bit, the sign bit, must be added to the codeword. The result is the sign-magnitude BCD code.

2.7.5 Specifications of Data Converters

With measurement data acquisition systems, it is not only the testing of the individual components and the basic circuit arrangement that is of interest. Only the evaluation of the entire system with its numerous possible sources of error allows a conclusion to be drawn about the dynamic uncertainty of the system. Possible error sources are shown in Fig. 2.74. Since not all uncertainties of such a complex system can be covered by a single test, several different test procedures have been developed.

The beta frequency test can be used to detect various digitization errors in a measurement data acquisition system. For this purpose, the signal frequency and sampling frequency are selected to be almost identical, and a sample is taken from each sine wave of the test signal. To achieve that all possible quantization steps can occur in the sampled signal, the frequency difference must be chosen so large that a difference of <1 LSB can be expected between two consecutive samples. This allows the so-called "missing codes" of the sampling system to be determined.

Dynamic errors can also be detected with the envelope curve test. For this purpose, a sinusoidal signal is selected whose frequency corresponds to about half the sampling frequency. Since successive measured values have changing signs, this places a great dynamic demand on the system. The deviation of the signal frequency from twice the sampling frequency determines, for a given memory length, the displayed envelope frequency. Missing AD combinations can also be detected here.

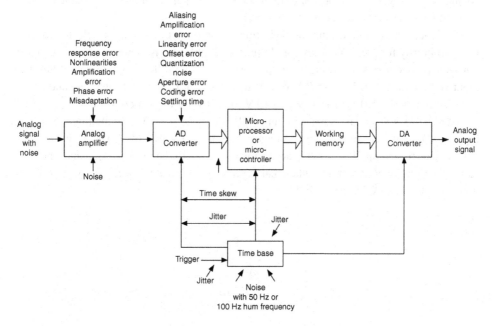

Fig. 2.74 Possible sources of error in a measurement data acquisition system

To determine the overall dynamic uncertainty of a system, a method is often used which determines the "effective bits". This method provides information about the effective resolution of a system. For this purpose, a sinusoidal voltage with a low harmonic distortion factor is applied to the DUT. For high-resolution systems (>10 Bit) the use of an additional low-pass filter is necessary. Afterward, an attempt is made to approximate the recorded data in the best possible way using a sine function (curve fitting). This sine function can be determined with the microprocessor or microcontroller. Then a comparison is made between the curve generated in the microprocessor or microcontroller, the recorded data, and the deviation is expressed by the effective value of the error function. This RMS value can be compared to the RMS value of the noise that an ideal AD converter, with the same number of bits as the data acquisition system under test, has.

The resolution of a converter is the smallest possible step (quantization interval Q) that can be distinguished from a converter. This step is also known as LSB. For a binary converter, the resolution can be determined by dividing the full input range by the number of quantization intervals ($FS/2^n$ with n number of bit positions). For a BCD converter, $logFS/2^d$ with d is the number of digits or decimal places. The value of FS is the full input voltage range (full scale). The resolution is a theoretical value and does not define the actual accuracy of a converter. A converter with resolution $1/2^{10}$ has a resolution of about 0.1% of the full input range, but its accuracy can still be only 0.5% due to other errors.

Some manufacturers specify the resolution in dB values. For this purpose, the approximation derived from the calculation of the quantization noise is used, i.e. that a dynamic range, i.e., a signal-to-noise ratio of approx. 6 dB can be achieved per bit of resolution. Again, this is only a theoretical value. Table 2.6 shows which theoretically possible resolution can be achieved for an *n-bit AD converter.*

The input voltage ranges and the definition of the output code can be selected essentially arbitrarily for the AD conversion. In practical applications, however, there are input ranges and code forms determined for AD converters. The following values apply to the input voltage ranges: 0 V to +5.0 V or 0 V to +10.0 V for unipolar converters and from −2.5 V to +2.5 V, −5.0 V to +5.0 V or −10 V to +10 V for bipolar converters.

Example: A sinusoidal signal with an amplitude of $U_e = 10$ V is to be digitized by a bipolar AD converter with an input voltage range of −10 V to +10 V. The 12-bit format corresponds to several $2^{12} − 4096$ quantization stages. With the measuring span of the AD converter from 20 V, the input voltage can be resolved at $1/4096$ or represented accordingly with 0.024%. With the formula

$$\begin{aligned}
DR &= 20 \cdot \lg 2^n \\
&= 20 \cdot \lg 2^{12} \\
&= 20 \cdot 12 \cdot \lg 2 \\
&= 20 \cdot 0.301 \\
&= 72.25 \text{dB}
\end{aligned}$$

Table 2.6 Resolution of an *n-bit AD converter* that converts to binary code

Bit	$1/2^n$ (parliamentary group)	$1/2^n$ (decimal)	ppm
MSB	1/2	0.5	50
2	1/4	0.25	25
3	1/8	0.125	12.5
4	1/16	0.0625	6.25
5	1/32	0.03125	3.125
6	1/64	0.015625	1.6
7	1/128	0.0078125	0.8
8	1/256	0.00390625	0.4
9	1/512	0.001953125	0.2
10	1/1024	0.0009765625	0.1
11	1/2048	0.00048828125	0.05
12	1/4096	0.000244140625	0.024
13	1/8192	0.000122703125	0.012
14	1/16.384	0.00006103515625	0.006
15	1/32.768	0.000030517578125	0.003
16	1/65.536	0.000015287890625	0.0015
17	1/131.072	0.00000762939453125	0.0008
18	1/262.144	0.00000381469726625	0.0004
19	1/524.288	0.0000019073486328125	0.0002
20	1/1.048.576	0.00000095367431640625	0.0001

This gives a dynamic of 72.25 dB. For the signal with an input voltage of 10 V, the numerical value 20 V/4096 = 4.88 mV is obtained as the smallest quantization unit.

The relationship between the resolution of an AD converter and the resulting number of quantization steps is shown in Fig. 2.75. Thus, already when using an analog-to-digital converter with a resolution of 10 Bit, an accuracy of <0.1% corresponding to a dynamic of >60 dB can be achieved. These values are often better than the specifications of numerous sensors, which means that the technical and thus costly effort for the anti-aliasing filter with the subsequent digitizing circuit can still be kept within acceptable limits. Many measuring systems with digital output on the market are therefore satisfied with this acceptable accuracy.

2.7.6 Relative Accuracy of Transducer Systems

The characteristics of a converter system (AD converter at the input or DA converter at the output, including the peripherals) are essentially determined by several factors, which are explained in more detail below.

When considering the function of a transducer system, a distinction is made between ideal and real components. Since there are no ideal components in practice, deviations

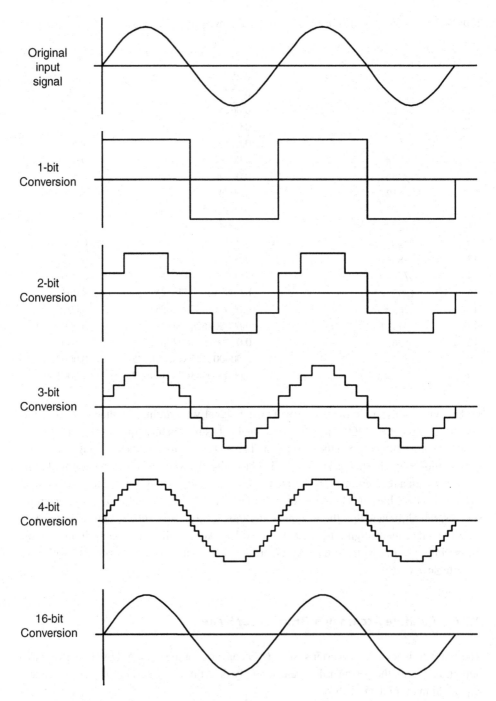

Fig. 2.75 Relationship between the resolution of a sinusoidal AC voltage in bit positions (resolution) and the resulting number of quantization steps

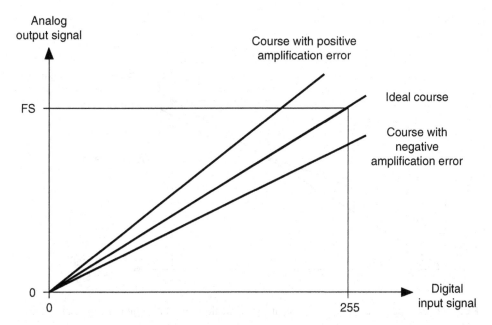

Fig. 2.76 Characteristic output curve of an uncalibrated transducer system. The digital input signal in 8-bit formal is converted into an analog output voltage between 0 V and the maximum output voltage "FS"

from the ideal behavior occur, caused by tolerances of the individual components and by different temperature coefficients. In practice, therefore, deviations from the ideal characteristic curve of a transducer occur.

Figure 2.76 shows such a characteristic curve, where FS (full scale) is the maximum output value of a transducer. It should be noted that the output characteristic curve shown in Fig. 2.76 and the following figures as a continuous function is in reality a finely graded step function with 255 (8-bit converter), 1023 (10-bit converter), 4095 (12-bit converter) steps, etc. Deviations of the characteristic output curve from the ideal curve can be classified as follows:

- Offset error: The offset error is the output signal value that is present when digital word 0 is applied to the input. As shown in Fig. 2.77, the zero point deviation causes a parallel shift of the characteristic output curve.
- Gain error: The gain error describes the deviation of the output signal from the setpoint when the highest digital word (255, 1023, 4095, etc.) is present at the input. As shown in Fig. 2.78, the gain error causes the characteristic output curve to rotate around zero.
- Linearity error: The linearity error describes the fact that the output characteristic curve is generally not a straight line, but has a somewhat wavy or curved shape, as shown in Fig. 2.76.

When examining the accuracy of a transducer, all sources of error must be considered and a distinction must be made between absolute and relative accuracy. A mere indication

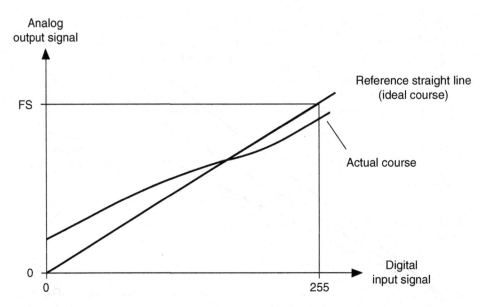

Fig. 2.77 Zero point deviation of the output characteristic curve of an 8-bit DA converter system

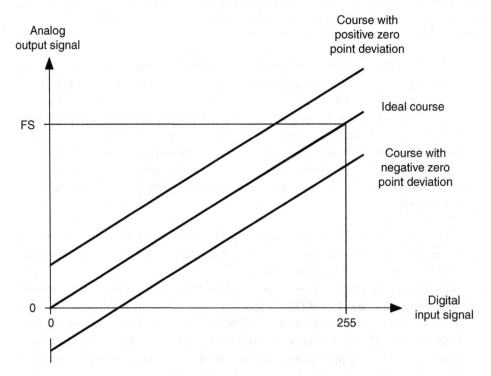

Fig. 2.78 Ideal characteristic curve and characteristic curves with possible gain errors in an 8-bit DA converter system

of the resolution, which influences the error limits of the conversion result, is therefore not sufficient to determine the accuracy of a transducer. The relative accuracy can be defined as linearity, whereby a distinction is made between the integral and the differential linearity error. Since most transducer systems offer the possibility to adjust the gain and offset error to zero by external trimmers, the relative accuracy can be defined as the linearity, whereby a distinction is made between the integral and differential linearity error. These two errors are not taken into account when estimating the accuracy—assuming careful adjustment and corresponding long-term stability of the transducer.

2.7.7   Absolute Accuracy of Transducers

The absolute accuracy of a transducer is defined as the percentage deviation of the maximum real output voltage FS (FSR range-1/2 LSB) from the specified measuring span FSR (maximum measuring voltage minus minimum measuring voltage). Sometimes an accuracy error is also specified in the accuracy specification. An accuracy error of 1% corresponds to the absolute accuracy of 99%. The absolute accuracy is determined by the three individual sources of error, such as the inherent quantization error (this is entered as ±LSB error), the errors caused by non-ideal circuit components of the transducer design and the conversion error derived later. Since the absolute accuracy is also influenced by temperature drift and long-term stability, the error for the defined ranges must also be specified when specifying the accuracy of a transducer.

Absolute accuracy can be derived from the parameters. The absolute accuracy is mainly determined by the conversion process and is also dependent on the two factors conversion time and quality of the converter. Relative accuracy is the maximum deviation of the output characteristic curve of the converter system from the ideal characteristic curve, which, according to Fig. 2.79, connects the zero point with the output signal that occurs when the digital word 255, 1023, 4095, etc. is entered.

The term relative accuracy covers all errors that occur due to non-linearities of the converter during conversion. For DA converters, the direct specification of the non-linearity is preferred, whereas these errors are simply added up under the term relative accuracy for AD or DA converters. The relative accuracy of conversion can only be increased if a converter with higher accuracy is used. Non-linearity errors of a converter cannot be adjusted.

Another possibility on the hardware side is the use of digital error correction circuits or on the software side, the use of special error algorithms provided a process computer is integrated with a system. If, for example, higher accuracy requirements are placed on an 8-bit conversion, this condition can be met by using a 12-bit converter in which only the first 8-bit digits calculated by the MSB are used. If the 12-bit converter has a non-linearity of ±1/2 LSB, the non-linearity error is reduced to 1/32 LSB if the first 8-bit digits are used exclusively. When estimating the relative accuracy, a distinction is made between integral and differential nonlinearity.

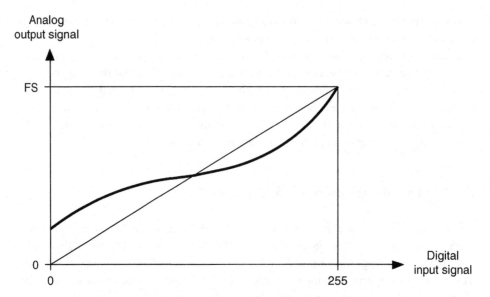

Fig. 2.79   Real (*thick*) and the ideal characteristic curve of an 8-bit DA converter

The concept of linearity corresponds essentially to the definition of relative accuracy. However, the reference straight line is not predetermined but can be designed so that the smallest deviations from the respective transducer characteristic curve occur.

The differential linearity describes the error in the output signal step height, relative to the setpoint step height, caused by a change at the digital input of 1 LSB. Figure 2.80 shows the definition of differential linearity in a transducer system.

The term differential nonlinearity covers the amount of deviation of each quantization result, i.e., any possible output code that does not correspond to its theoretical ideal value. In other words, differential nonlinearity is the analog difference between two adjacent codes from their ideal value ($FSR/2^n$ = 1 LSB). If the differential non-linearity value of ±1/2 LSB is specified for an AD converter, the value of each minimum quantization step, relative to its transfer function, is between 1/2 and 3/2 LSB, i.e., each analog step is 1 LSB ± 1/2 LSB.

The behavior of a DA-converter is called monotonic if the output signal increases or remains the same when the digital value increases. Figure 2.80 shows a monotonous and Fig. 2.81 a non-monotonous function. Monotony is present if the differential linearity is less than 1 LSB.

A converter should not only convert the digital input signal into an analog value as accurately as possible, it should also do this with minimum delay. The delay can be described by the settling time. This term includes the time from the application of the digital codeword until the output signal does not exceed specified limits, usually ±1/2 LSB (Fig. 2.82). The value usually refers to specific test conditions and may depend on the output circuitry. This applies to both AD converters and DA circuits.

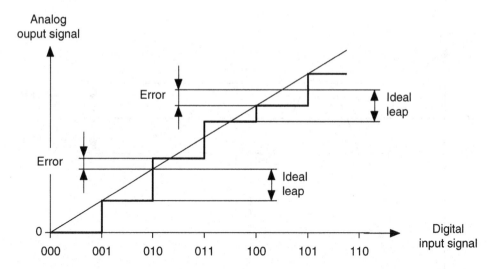

Fig. 2.80 Definition of differential linearity for a monotonic transducer system

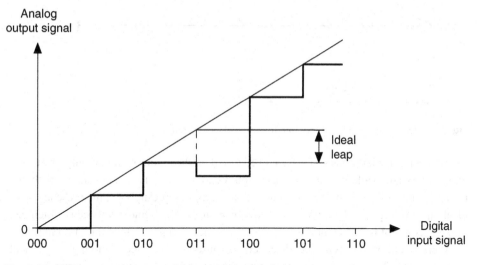

Fig. 2.81 Non-monotonic course of the output characteristic curve

In the case of converter systems, the corresponding stability specifications refer to the analog or digital output signal. A distinction is made between the dependency of this signal on the temperature and the operating voltage or their changes. The influence of the temperature is specified in ppm/K, the influence of the operating voltage in mV/V or µA/V, in each case referred to the full-scale value. The quotient of the percentage change of the full-scale value and the percentage change of the operating voltage can also be used as a stability characteristic.

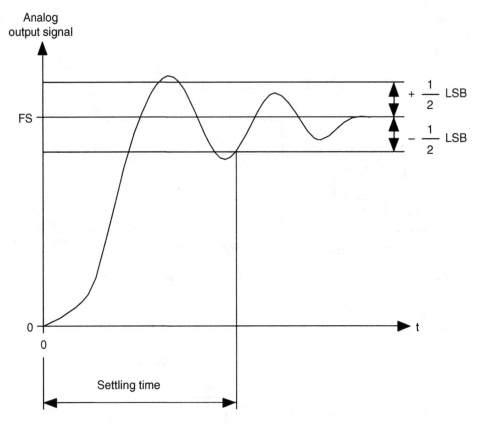

Fig. 2.82 Definition of the settling time for a transducer system

All error considerations so far have been based on the assumption that only ideal sampling pulses are available for AD conversion, i.e., the pulses have a constant length and their time intervals are always the same. This assumption presupposes the presence of an ideal input voltage, furthermore, no interference on the clock lines may occur which distorts the edges of the clock pulses and thus shifts the starting points of the sampling.

Aperture uncertainty is used to detect errors caused by non-ideal sampling (in terms of time intervals). The effects of sampling jitter can be understood as noise superimposed on the ideal conversion result.

According to Fig. 2.83, the following approach can be used to derive the error:

$$F\left[t+\Delta T\left(t\right)\right]=f\left(t\right)+f^{*}\left(t\right)\Delta T\left(t\right)+\ldots$$

With

$$f^{*}\left(t\right)\cdot\Delta T\left(t\right)$$

the error caused by the sample jitter can be described. The mean value of the noise voltage is then

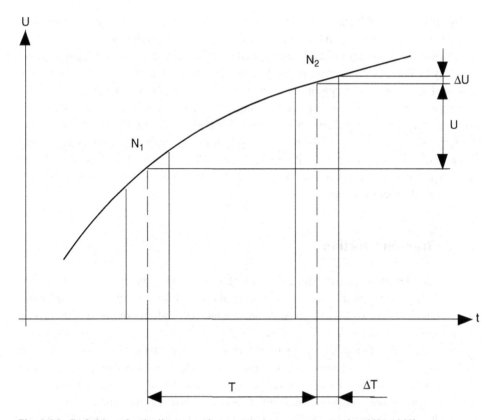

Fig. 2.83 Definitions for the jitter error between two measurement points *N1* and *N2*

$$\left[f^{*}(t)\cdot\Delta T(t)\right]^{2}=\left[f^{*}(t)\right]^{2}\cdot\left[\Delta T(t)\right]^{2}$$

The effective value of the noise voltage is the square root. If a sine voltage with frequency f and amplitude $\hat{u}$ (peak value) is applied to the input of the AD converter, the effective sampling jitter error can be calculated with the equation

$$E_{\text{Jitter(RMS)}}=2\cdot\pi\cdot\hat{u}\cdot f\left(\Delta T_{\text{RMS}}\right).$$

2.8   Digital to Analog Converter

Every PC system or process computer works digitally, i.e., with the information values 0 and 1. To communicate with the analog outside world, DA converters are required which convert the digital values of the computer into analog output variables. The areas of application for DA converters range from video and audio applications, automatic test systems, or digitally controlled loops to process control. Besides, DA converters are the main components of some important AD converter families in practical applications.

DA converters convert a discrete signal $s(n)$ into a proportional analog signal $s(t)$. Since $s(n)$ only changes to T at certain points in time, the analog output signal also changes at these points in time. The amplitude at the output of a DA-converter has a rectangular shape, i.e., there is a time-continuous, but value discrete function. The signal $s(t)$ contains because of its rectangular shape unwanted parts of higher frequencies, which result from the harmonic frequency values of the conversion frequency $f_w = 1/T$. To eliminate these distortions, a DA-converter must always be followed by a passive low-pass filter of first order, which has a cut-off frequency of $f_0 = f_w/2$. This RC combination (resistor and capacitor) is called a reconstruction filter.

2.8.1 Transfer Function

Several code forms are possible for DA-converters. In practice, one works either with the natural binary code with the binary coded decimal code (BCD) or with their complementary versions. Each code has its specific advantages and disadvantages in certain applications.

The transfer function of an ideal 3-bit DA converter is shown in Fig. 2.84. Each input code generates its output value, usually a direct or negated current value. In cases where an operational amplifier for a current-voltage converter is already present in the device, a corresponding output voltage is generated. If there is a current output, the conversion into an output voltage is either done in the simplest case via a resistor, or an operational amplifier is controlled, which works as a current-voltage converter. The entire output range comprises 2^n values, including zero. These output values show a 1:1 match with the input, which is not the case with AD converters. Since 1994, more and more DA converters with a direct voltage output have been found in practice. This allows us to save the external operational amplifier.

There are many different ways to realize DA-converters. In practice, however, only a few of these are used in circuit design. With parallel DA-converters, the digital data are simultaneously "parallel" to the converter. With serial DA-converters, the analog output value is only obtained when all digital values have been written sequentially.

The DA converter also has 2^n output states with $2^n - 1$ transition points between the individual states. The value Q is the difference between the analog amount and the transition points. For the converter, Q represents the smallest analog difference amount that the converter can resolve, i.e., Q is the resolution of the converter expressed for the smallest analog amount.

Usually, the resolution is given in bits for converters, which indicate the number of possible states of a converter. A DA converter with an 8-bit resolution produces 256 possible analog output stages, whereas a 12-bit converter produces 4096 values. In the case of an ideal converter, the value Q has the same value over the entire range of the transfer

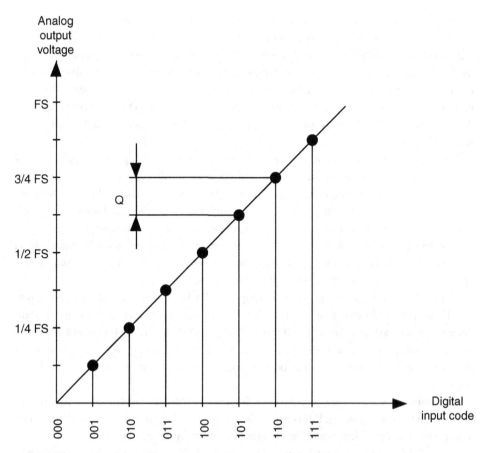

Fig. 2.84 Transfer function of an ideal 3-bit DA converter with the value FS (full scale) and the quantization error Q

function. This value is represented as $Q = FSR/2^n$, where FSR indicates the measuring span, i.e. the difference between the minimum and maximum measuring range (FS).

For example, if a transducer is operated in the unipolar range between 0 and +10 V or the bipolar range from −5 V to +5 V, the FSR value is, in any case, $U = 10$ V. The FS value, on the other hand, is 5 V in the unipolar range and 10 V in the bipolar range.

The factor Q is also used to denote the LSB value since this represents the smallest code change that a DA converter can produce. The last or smallest bit in the code changes from 0 to 1 or from 1 to 0.

It should be noted that the output value never quite reaches the maximum measuring range (FS) when using the transfer functions of a transducer. This results from the fact that the maximum measuring range is a nominal value that is independent of the resolution of the converter. For example, a DA-converter has an output range from 0 to +10 V, where 10 V represents the nominal scale end range (= maximum measuring range). For example, if the DA converter has an 8-bit resolution, the maximum output value is $255/256 \cdot 10$ V = 9.961 V. If, on the other hand, a 12-bit converter is used, the maximum output voltage is $4095/4096 \cdot 10$ V = 9.9976 V. In both cases, the maximum output value therefore only reaches 1 Bit less than that indicated by the nominal output voltage. This is because the analog zero value already represents one of the two converter states. There are therefore only $2^n - 1$ steps above the zero value for both AD and DA converters. To reach the scale range, $2^n + 1$ states are necessary, which means that an additional code bit is required.

For the sake of simplicity, data converters are therefore always given in specifications with their nominal range instead of their real achievable full-scale value. In the transfer function of Fig. 2.84, a straight line is drawn through the output values of the DA-converter. In an ideal converter, this line runs exactly through the zero points and through the full-scale value. Table 2.7 shows the most important characteristics of data converters.

Everyone, even the ideal converter, has an unavoidable error, namely quantization uncertainty or quantization noise. Since a data converter cannot detect an analog difference of $<Q$, its output is subject to an error of $\pm Q/2$ at all points.

Figure 2.85 shows a circuit for examining an 8-bit DA converter with current output. The IDAC8 has eight data inputs, which are driven by a bit pattern generator in the hexadecimal number system. The controller must be set to "Cycle". If the bit pattern generator does not work properly, the settings of the definition are missing, because this must be set to "Count up". The settings must be readjusted by the encoders each time Multisim is called.

The staircase voltage runs in a positive direction because the reference current source is connected to the I_{ref-} input. If this is connected to the I_{ref+} input, a negative staircase voltage results. The operational amplifier operates in inverting mode.

Figure 2.86 shows a circuit for examining an 8-bit DA converter with voltage output. The VDAC8 has eight data inputs, which are driven by a bit pattern generator in the hexadecimal number system. The control must be set to "cycle". If the bit pattern generator is not working properly, the definition is missing, because it must be set to "Count up".

Table 2.7 Characteristic features for data converters

Resolution n	States 2^n	Binary weighting 2^{-n}	LSB for FS = 10 V	Signal to noise ratio dB	Dynamic range dB	Maximum output for FS = 10 V
4	16	0.0625	625 mV	34.9	24.1	9.3750 V
6	64	0.0156	156 mV	46.9	36.1	9.8440 V
8	256	0.00391	39.1 mV	58.9	48.2	9.9609 V
10	1024	0.000977	9.76 mV	71.0	60.2	9.9902 V
12	4096	0.000244	2.44 mV	83.0	72.2	9.9976 V
14	16.383	0.000061	610 μV	95.1	84.3	9.9994 V
20	65.536	0.0000153	153 μV	107.1	96.3	9.9998 V

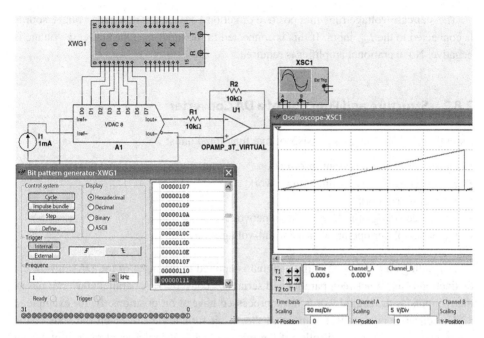

Fig. 2.85 Investigation of an 8-bit DA converter with current output

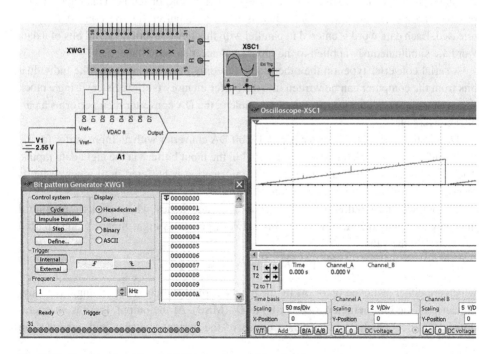

Fig. 2.86 Examination of an 8-bit DA converter with voltage output

The staircase voltage runs in a positive direction because the reference voltage source is connected to the I_{ref+} input. If this is connected to the input I_{ref-}, the staircase voltage is negative. No operational amplifier is required.

2.8.2   Structure and Function of a DA Converter

Essentially, a DA converter consists of five functional units:

- Buffer for the digital input information
- For the control of the evaluation network
- Evaluation network
- Reference voltage for the evaluation network
- Output amplifier operating as a current-voltage converter

Depending on the way the digital signal is applied to the input of the DA converter, a distinction is made between parallel and serial DA converters. A parallel converter has as many inputs as the digital words to be processed have at bit positions. If, for example, a 12-bit converter is present, it can adopt the 12-bit format via the 16-bit data bus of the microprocessor or microcontroller. If the microprocessor or microcontroller only has an 8-bit data bus, the 12-bit converter first takes over the lower byte, and in a second write operation the upper byte, which is then present as a nibble or tetrad. This data transfer process can also be reversed. Only then is the 12-bit buffer for the converter output released. Each data word is entered in parallel with this DA-converter, i.e., all bits of a data word are simultaneously applied to the inputs and accepted simultaneously.

A serial converter type, on the other hand, has only one data input and the individual bits from the computer can be written into the buffer memory (shift register) using a clock line. Once the serial data transmission is complete, the DA converter then performs a parallel conversion,

The circuit in Fig. 2.87 shows a parallel 8-bit DA converter with an input buffer for the digital information, which is stored in parallel in the input buffer via the eight data inputs. The LE input is used for data transfer. This information is buffered until a new value is received from the microprocessor or microcontroller.

The outputs of the buffer are connected to current switches, which are used to control the evaluation network. In this case, this network generates eight binary graded partial currents with the values

$$128 : 64 : 32 : 16 : 8 : 4 : 2 : 1$$

which are either switched on or off by the binary current switches, depending on the significance of the bit coefficients D_0 (LSB) to D_7 (MSB). At the output, there is an operational amplifier which combines the outputs of the current switches and generates an

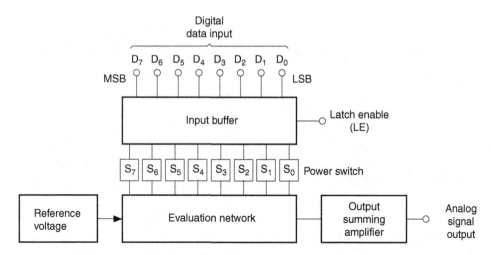

Fig. 2.87 Structure of a parallel 8-bit DA converter with an interface for a microprocessor

analog output signal. Figure 2.88 shows a 4-bit DA converter with the 1248 resistor network.

Important in this circuit is the reference unit, which generates either a voltage or a current for the evaluation network. This reference unit does not necessarily have to be present in the DA-converter, but can also be added externally.

With an ideal DA-converter, there is a linear 1:1 dependency between the digital input information and the analog output value.

2.8.3   R2R-DA Converter

A very well-known technique in DA converters is the R2R conductor network. As can be seen in Fig. 2.89, this network consists of series resistors with the value R, and the shunt resistors have a value of $2R$. The open end of the $2R$ resistor is connected via a single-pole switch to either ground or the current sum point of the downstream operational amplifier. The single-pole changeover switch is realized with the corresponding analog switches.

The principle of the R2R network is based on the binary division of the current through the network. A closer examination of the resistor network shows that viewed from point A to the right, a measured value from $2R$ is obtained. Thus the reference voltage detects a network resistance of R.

At the reference voltage input, the current divides into two equal parts because it detects the same resistance value in each direction. Likewise, the currents flowing to the right are divided in the same ratio in the subsequent resistance nodes. The results are binary weighted currents flowing through all 2R resistors of the network. The digitally controlled switches then carry these currents either to the summation point or to the ground.

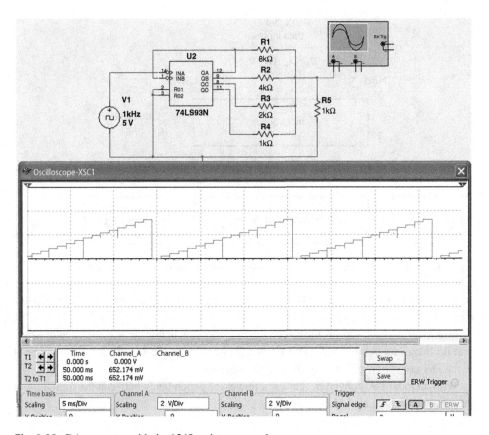

Fig. 2.88 DA converter with the 1248 resistor network

Assuming that all inputs have a 1 signal, as shown in Fig. 2.89, all current will flow to the summing point. As can be seen in the figure, the sum current is fed to an operational amplifier, which converts the current into a corresponding output voltage. The output current is calculated from

$$I_a = \frac{U_{ref}}{R}\left(\frac{1}{2}+\frac{1}{4}+\frac{1}{8}+\ldots+\frac{1}{2^n}\right)$$

which is a binary control. This means that for the sum of all currents

$$I_a = \frac{U_{ref}}{R}\left(1-2^{-n}\right)$$

The expression 2^{-n} represents the physical part of the current flowing through the terminating resistor $2R$ located at the extreme right end. The advantage of the R2R resistor network technology is that only two different resistance values need to be trimmed, from which a good temperature behavior can be derived. Besides, relatively low-ohm resistors

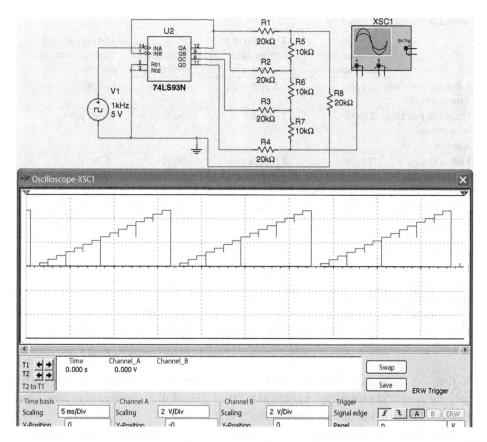

Fig. 2.89 DA converter with the R2R conductor network

are used for fast applications. For high-resolution DA-converters, very precisely working laser-trimmed thin-film resistor networks are suitable.

2.8.4   DA-Converter with External Resistors

The DA-converters, which one wants to build discretely, differ among other things by the use of different resistor materials. However, the resistors are of crucial importance for data conversion components. For example, a typical digital-to-analog converter is constructed with a binary-weighted R2R resistor network. Decisive for the linearity and differential linearity are the resistance ratios among each other. The tolerance of the MSB resistor is most critical here. The current through this branch must maintain the tolerance of $\pm 1/2$ LSB. If this is not the case, the converter loses accuracy (monotony, differential linearity, and linearity). An n-bit converter would deteriorate to a $(n - 1)$ bit converter.

Table 2.8 Stability of resistors

	Absolute temperature coefficient	The ratio of the temperature coefficients	Typical drift per year	The typical drift of the resistance ratios per year
Laboratory wound wire resistors	1 ppm	0.5 ppm	2–5 ppm	1–5 ppm
Industrial precision wire wound resistors	2 ppm	1–2 ppm	15–50 ppm	10–30 ppm
Film resistors	3 ppm	1.5 ppm	25 ppm	15–40 ppm
Printed special thin-film resistors	5–15 ppm	3–5 ppm	25–50 ppm	25–50 ppm
Thin-film resistors	20–60 ppm	2–6 ppm	200–400 ppm	100–400 ppm
Laser adjusted miniature thin film resistors	20–60 ppm	3–10 ppm	200–600 ppm	200–600 ppm
Discrete thin film resistors (RN55E)	25 ppm	10 ppm	300–1500 ppm	500–2000 ppm
Balanced thin film resistor networks	50–100 ppm	5–50 ppm	500–1000 ppm	200–2000 ppm
Carbon film resistors	1000–2000 ppm	500–1000 ppm	20,000 ppm	20,000 ppm

Table 2.8 shows a comparison of usable resistors with the absolute temperature coefficient, the ratio of the temperature coefficients, the typical drift of resistors after 1 year, and the typical drift of the resistance ratios after 1 year.

The current possible production of semiconductor ICs without laser alignment is at 12-bit resolution. With laser alignment of the resistor networks, accuracies of up to 16-bit resolution can be achieved. The disadvantage of laser alignment, however, is that no information about the long-term drifts can be given. With discrete resistors, these drifts are known and after a "burn-in" they allow a ratio drift from 10 ppm to 30 ppm. This value is sufficient for a 16-bit accuracy (0.0015%) after 1 year.

As can be seen from Table 2.8, only the high-quality resistor materials up to a long-term drift of 50 ppm per year meet the requirements of 14-bit to 16-bit converters. The resistor materials listed in the lower part of this table do not meet the requirements of a 12-bit converter without additional adjustment.

Figure 2.90 shows a circuit of the DA converter with an 8421 resistor network. The DA converter circuit design combines the advantages of weighted current source technology with the 8421 resistor network. This scheme does not use large switching transistor currents to operate the individual nodes.

Figure 2.90 shows voltage dips. The cause of the voltage peaks is due to small time differences between the switching on and off of switching transistors and in this case of the asynchronous counter. An example is a 1 LSB increase in the 1/2 FS range. If the MSB of the second code is switched on faster than the bits in the first code are switched off,

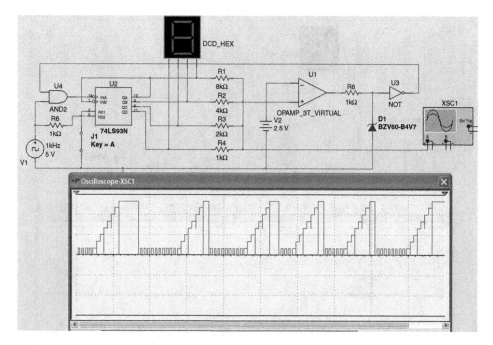

Fig. 2.90   DA converter with 8421 resistor network

practically all bits are switched on for a short time, resulting in a voltage peak of almost 1/2 FS. This peak can then be detected on the oscilloscope as interference. The glitches can be eliminated with a capacitor from 1 nF at the output.

The counter 74193 is an asynchronous counter and in counting mode, there are intermediate states that are fully effective here. If the synchronous counter 74163 is used instead of the 7493/74193, no intermediate states and thus no glitches occur in counter mode.

If the 7493 TTL counter module is used in Fig. 2.91, the asynchronous mode of operation again results in intermediate states and this can be seen in the dips. If the synchronous counter 74163 is used instead of the 7493, there are no intermediate readings and thus no glitches. The advantage of the R2R resistor network are the identical currents so that all emitter resistors are of the same size, and thus largely identical switching speeds can be achieved.

A data register is connected in series to the input of the DA converter for intermediate storage of the digital information and the output is connected to a sample and hold amplifier. If a new digital word is now read in via the register, the sample and hold amplifier simultaneously switches to "Hold", i.e., the output circuit is interrupted and is held at the last analog voltage value. After the DA-converter has settled to the new value and the glitches have subsided, the sample and hold unit can switch back to "sample". This closes the circuit and the new output value is available without falsification.

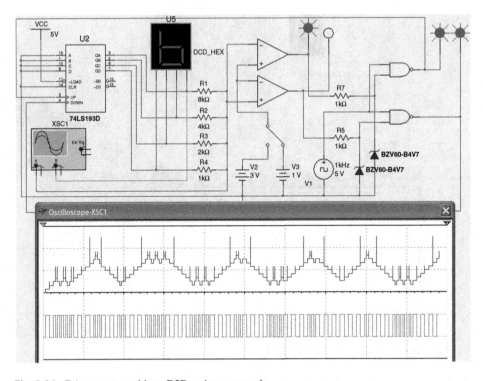

Fig. 2.91 DA converter with an R2R resistor network

2.9   Analog to Digital Converter

Analog-to-digital converters operate according to very different conversion methods. Similar to the DA-converters, however, only a few procedures are used in practice. The choice of the method is primarily determined by the resolution and the conversion speed.

There are numerous known methods for digitizing an analog input voltage. From the fact that a special AD module has been developed for each conversion method, it can be concluded that each AD method can be used advantageously under certain application conditions. In addition to the fundamental errors arising during conversion, each conversion method also contains system-related errors. For the user of AD converters, some basic knowledge of the different conversion methods is therefore very advantageous.

2.9.1   AD Converter According to the Counting Method

With an AD converter, which works according to the counting method, the pulse sequence of a clock generator is switched to a counter. The outputs of the counter are connected to a DA converter. With each clock pulse, the output voltage of the DA converter increases by 1 LSB. The output voltage at the DA converter is compared with the measurement or input voltage via a comparator, i.e. an operational amplifier with high open-circuit amplification. If the output voltage of the DA converter has reached the value of the measurement or input voltage, the comparator switches and blocks the clock generator. Since the evaluation of the pulses is performed, the measured value is shown in the seven-segment display.

An AD converter, which operates according to the counting method, requires four functional units. The comparator compares the measuring voltage U_e with the output voltage from the DA converter. If the output voltage of the DA converter is lower than the measuring voltage, the output of the comparator has a 1 signal and the clock generator can operate. The clock generator generates a certain frequency which drives the up-counter. The value of the counter increases by +1 for each clock pulse. The value of the counter determines the input value of the DA converter and thus its output voltage via the eight lines. The counter output represents the converted digital word, which can be queried by a computer via a parallel interface. If the output voltage U_a reaches the level of the measuring voltage, the comparator switches to 0-signal at its output and thus stops the clock generator.

The AD converter of Fig. 2.92 can be controlled, started, or triggered with a clear signal. With the clear signal, the up-counter is reset to 0. The DA converter thus has a voltage of $U_v = 0$ V at its output and the output of the comparator switches to 1 signal. This signal

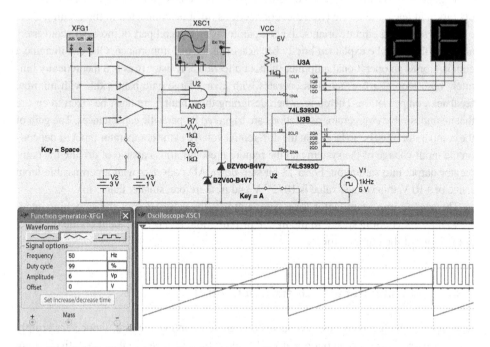

Fig. 2.92 Circuit of an analog-to-digital converter with a pulse diagram according to the counter or step conversion method. The output of the up-counter represents the converted digital word

is used to control the clock generator, which starts the up-counting from counter reading 0. The counter is stopped when the condition $U_e = U_v$ is fulfilled.

This circuit variant has advantages and disadvantages. One of the advantages is the realization of a peak value AD converter. If the measuring voltage increases, the up-counter starts up and tries to reach the amplitude of the input voltage via the downstream DA-converter. If the measuring voltage decreases, no change occurs. The disadvantage is the long time needed for a complete conversion after the start signal.

This TTL module 74393 contains a twofold and an eightfold divider. The asynchronous counter 74393 consists of four flip-flops which are internally connected in such a way that one counter up to two and one counter up to eight are created. All flip-flops have a common reset line which can be used to reset them at any time (pin 2 and pin 3 to 1 signal). Flipflop A is not internally connected to the other stages, which allows different counting sequences:

(a) Counting up to 16: For this purpose, the output Q_A is connected to the clock input IN_B. The input frequency is fed to connector IN_A and the output frequency is taken from Q_D. The block counts in binary code up to 16 (0–15) and resets itself to zero at the 16th pulse.

(b) Counting to 2 and counting to 8: Here the flipflop A is used as a divider 2:1 and the flipflops B, C, and D as dividers 8:1.

Triggering is always on the negative edge of the clock pulse. For normal counting operation at least one of the two reset connections must be connected to the ground.

In the following circuit variants, the comparator is an integral part of most A/D converters and shall therefore be explained briefly because of its special importance. One can imagine a comparator as an operational amplifier without a feedback resistor (this is a theoretically infinitely high loop gain), which thus operates with a very large gain bandwidth with minimal feedback compensation. Therefore, when designing the circuit, care must be taken to ensure that no undesirable comparator oscillations are triggered by parasitic capacitances. The gain of the comparator must be so large that the differential voltage at the comparator input (it depends on the input voltage of the system and the required resolution) is capable of driving the comparator output into saturation. For a 12-bit step-down AD converter with a permissible input range of +10 V, this voltage value is $10/2^{12}$ V and is, therefore, smaller than 1 mV.

The second important factor for a comparator is the estimation of the switching accuracy and the switching speed. In the first approximation, this is dependent on the overload at the input of the comparator.

2.9.2 AD Converter with Overshoot Control

Converters, which are constructed according to the counting method, work very simply and can be realized very economically. However, the conversion takes a relatively long time, because the counter is set to zero before the start. Figure 2.93 shows an improvement in this technique.

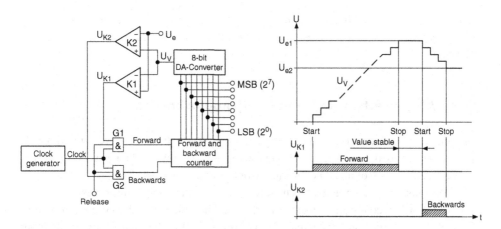

Fig. 2.93 Analog-to-digital converter with tracking control (tracking method) using an up/down counter

For the realization of an AD-converter with tracking control, two comparators are needed, which compare the output voltage of the DA-converter with the measuring voltage differently. From the two comparisons, the two AND gates can be controlled, which are connected to the common clock generator. Depending on whether the upper or lower comparator generates a 1 signal at the output, the forward or reverse input receives the counting pulses. If both voltages U_e and U_v are identical, both comparators generate a 0 signal for the two NAND gates, and the counter is blocked. If the measuring voltage changes in a positive or negative direction, the counter operates in forward or reverse mode.

The 74193 block contains a programmable synchronous 4-bit binary counter with separate clock inputs for up-counting and down-counting, as well as a clear input. For normal operation, connect the "Load" connector to the 1-signal and the "Clear" connector to the 0-signal. With each 01 transition (positive edge) at the clock input UP, the counter advances one step upwards. With each positive edge of the clock at the "Down" input, the counter goes down. The respective other clock input must be set to 1 signal. For programming, the required number is applied in binary code to the inputs P0 to P3 and the "Load" input is briefly set to 0 signal.

To clear the counter, briefly set the "Clear" to 1 signal. The clearing process is independent of the clock.

When counting up, the carry output pin 12 gives a negative pulse when 15 is reached. When counting down, a short negative pulse is generated when 0 is reached at output 13. For multi-digit counters, connect pin 13 (down carry) to the clock input "Clock-Down" of the next stage and pin 14 (up carry) to the clock input "Clock-Up" of the following stage.

The obvious advantage of a "tracking" converter is that it continuously follows the input signal and provides constantly updated digital data. However, the input signal must not change too quickly. Higher-frequency input signals can be measured if fast operational amplifiers are used which are specially designed for comparator operation. If the input signal changes only slightly, this method is very fast in practical application. By applying a digital signal to the counter, this circuit allows the choice between overshoot and hold operation.

2.9.3 AD Converter with Step-by-Step Approximation

For transducers with a medium to very high conversion rate, the successive approximation or stepwise approximation method, also known as the weighing method, is very important. In practice, over 90% of all AD converters work according to this principle.

Just like counting technology, this method belongs to the group of feedback systems. In these cases, a DA converter is located in the feedback loop of a digital control circuit, which changes its state until its output voltage corresponds to the value of the analog input voltage.

In the case of stepwise approximation, the internal DA converter is controlled by an optimization logic (SAR unit or successive approximation register) in such a way that the conversion is completed in only n steps at n-bit resolution.

Figure 2.94 shows this procedure. The center of the circuit is the successive approximation register with an optimization logic for the control of the DA-converter. The method is also called the weighing method, because its function is comparable to weighing an unknown load using a balance, whose standard weights are applied in binary order, i.e. 1/2, 1/4, 1/8, 1/16 kg. The largest weight is placed in the pan first. If the scale does not tilt, the next largest weight is added. However, if the scale tilts, the last weight placed on

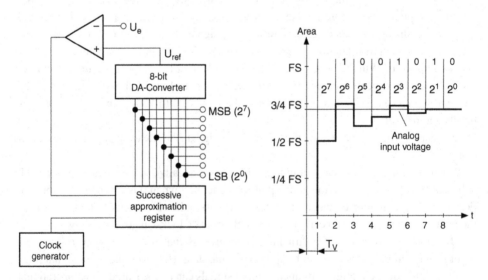

Fig. 2.94 Circuit and pulse diagram for an analog-to-digital converter operating according to the stepwise approximation

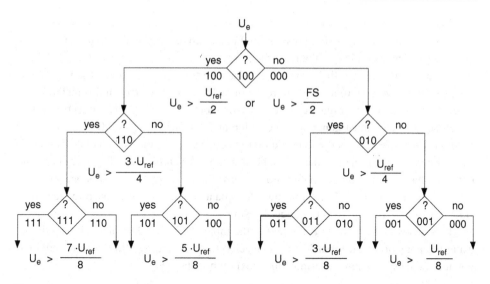

Fig. 2.95 Flowchart for a 3-bit converter operating according to the stepwise approximation

the pan is removed and the next smaller weight is placed on top. This procedure can be continued until the balance is in equilibrium or the smallest weight $(1/n$ kg) is placed on the pan. In the latter case, the standard weights resting on the pan represent the best possible approximation to the unknown weight. Figure 2.95 shows the flow chart of the stepwise approach.

The flowchart shows the operation of a 3-bit SAR unit. The successive approximation register first sets the MSB (most significant bit) after the start. For a 3-bit AD converter, this means that the value 100 is present at the output of the register. This value is at the DA converter and generates a corresponding comparison voltage U_v, which is compared with the measurement voltage U_e in the comparator. If the measurement stress is greater than the equivalent stress, the MSB position remains set in the SAR unit and a new comparison can be carried out. If the measuring stress is smaller than the equivalent stress, the MSB is reset to 0. After the first comparison, the register outputs the value 010 if $U_e < U_v$.

The converter and its subsequent stages repeat this process until the best possible approximation of the output voltage of the DA converter to the unknown measurement voltage has been achieved for a given resolution. The conversion time of the step-down converter can therefore be determined immediately. The time value is calculated at a resolution of n Bit from

$$T_u = n \cdot \frac{1}{f_T}$$

where f_T is the output frequency of the clock generator.

After n comparisons, the digital output of the SAR unit shows each bit position in its respective state and thus represents the coded binary word. A clock generator determines the time sequence. The effectiveness of this converter technology allows conversions in very short times at relatively high resolution. For example, it is possible to perform a complete 12-bit conversion in less than 100 ns.

Further advantages are the possibilities of a "short-cycle" operation, which results in even shorter conversion times without resolution. The source of error in this method is an inherent quantization error that occurs due to overshoot. If you have a 12-bit AD converter, the clock generator must generate three different frequencies (e.g. 1 MHz, 2 MHz, and 8 MHz). The 1 MHz frequency is needed for the conversion of the MSB and the following two. After that, the frequency increases, because now the amplitude difference of the output steps has become much smaller. For the last three bits of the conversion, the clock frequency can be increased again because the quantization units have been reduced considerably, so that overshoot is no longer possible.

The sinusoidal AC voltage is applied to the input V_{in} of Fig. 2.96 and can be adjusted using the potentiometer. Input V_{ref+} is connected to +5 V and V_{ref-} to 0 V (ground). The voltages on these two pins determine the maximum voltage V_{fs}

$$V_{fs} = V_{ref+} - V_{ref-}$$

To start the conversion, pin "SOC" (Start Of Conversion) has to be set to 1, which is done by the rectangle generator with f = 10 kHz, i.e., every 0.1 ms the AD converter performs a conversion. When a conversion is started, the output of pin "EOC" (End Of Conversion) is set to 0 signal, indicating that a conversion is taking place. When the conversion to 1 μs is completed, pin "EOC" is set to 1-signal again. The digital output signal is now available at pins D_0 to D_7. These tri-state output pins can be activated if pin "OE"

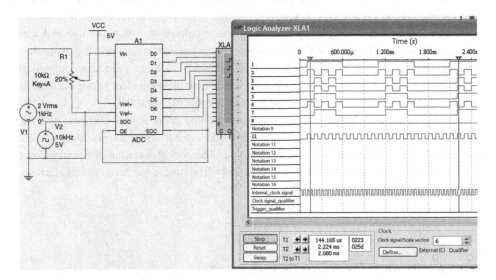

Fig. 2.96 Circuit and pulse diagram for a simulated 8-bit AD converter operating according to the successive approximation

(Output Enable) is set to 1-signal. The outputs of the AD-converter directly control the logic analyzer and you can see how the AD-converter works.

The digital output signal at the end of the conversion process is equivalent to the analog input signal. The discrete value corresponding to the quantization stage of the input signal is given by

$$\frac{\text{Input voltage} \cdot 256}{V_{\text{fs}}}$$

Note that the output signal described by this formula is not a continuous function of the input signal. The discrete value is then converted into the binary signal available at pins D_0 to D_7 and is given by

$$B_{\text{IN}} \left[\frac{\text{Input voltage} \cdot 256}{V_{\text{fs}}} \right]$$

2.9.4 Single-Slope AD Converter

Another class of AD converters, known as integrating types, operate on the principle of indirect conversion. The unknown input voltage is converted into a period. During this period, clock pulses are then evaluated by a counter. Based on this basic principle, there are several circuit variants, such as the single-slope, dual-slope and triple-slope method. Besides, another technique known as the batch balancing or quantified feedback method has become established.

Today, the single-slope AD converter of Fig. 2.97. is of practically no significance, but this technology forms the basis of the other converter types. As the name implies, this conversion method only uses a ramp for integration. This requires a generator that generates a precise sawtooth voltage and passes it on to two comparators. The upper comparator compares the sawtooth voltage with the ground (0 V), the lower comparator compares it with the measurement voltage.

The upper comparator starts the conversion cycle when the ramp voltage crosses the zero voltage line, while the lower comparator stops the cycle when the ramp voltage has reached the value of the input voltage. The time that elapses between these two switching points is registered in the counter modules utilizing a constant clock frequency and is proportional to the input voltage. With this conversion principle, however, the factor of a higher conversion speed must be expensively bought by a substantial improvement of the quality of the individual circuit elements. Long-term stability is of particular importance here so that this conversion method has become practically meaningless today.

The 74393 TTL module contains two completely separate binary counters with one reset input. Unlike the 74293, each of the two 4-bit binary counters does not have a separate B input. However, since all outputs are led out, the two counters can be used to

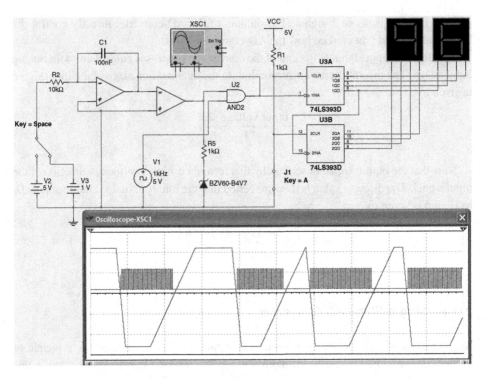

Fig. 2.97 Implementation of the single-slope AD converter

implement a wide range of divider options, namely 2:1, 4:1, 8:1, 16:1, 32:1, 64:1, 128:1 and 256:1. Each divider consists of four flip-flops and triggers at the 1 to 10 transition (negative edge) of the clock pulse. Each divider works in a 4-bit binary code. Besides, each divider can be set asynchronously to zero with its reset input by briefly connecting this input to a 1 signal. For normal counting operation, this input must be set to 0-signal.

2.9.5   Dual Slope AD Converter

In practice, the dual-slope or two-step procedure is used. Although this method is very slow, it has excellent linearity characteristics and the ability to completely suppress input noise or line voltage interference. Because of these features, digital voltmeters and multi-meters almost exclusively use integrating converters.

Figure 2.98 shows the circuit of a dual-slope AD converter, which essentially consists of six functional units. The integrator integrates the input voltage over time, resulting in a voltage rise in the positive or negative direction at the output, depending on the polarity of the input voltage. After a certain time, the control logic switches the input switch to the reference voltage U_{ref} and the capacitor C can discharge. At the same time, the control logic starts the counter. At the output of the integrator, the voltage rises continuously until

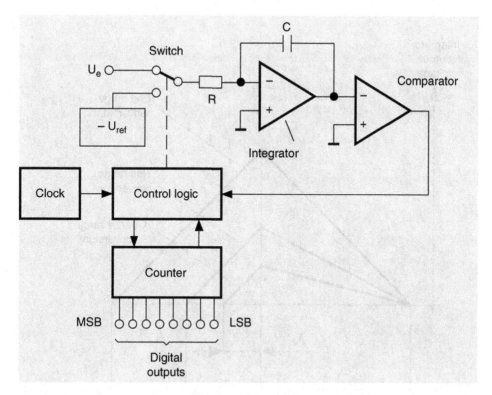

Fig. 2.98 Circuit of a dual-slope AD converter

the comparator detects a zero-crossing. The control logic evaluates this zero-crossing and the counter stops. The integration process is shown in Fig. 2.99.

The conversion starts as soon as the measuring voltage is applied to the input of the integrator. Via resistor R, capacitor C can be charged linearly in a positive or negative direction. When integration is started, the control logic activates the counter, which allows the fixed time t_1 in the counter to be determined. After the fixed time t_1, the integration of the input voltage is finished, and from this moment on, the second phase starts with the integration of the negative reference voltage, the duration of which is also counted. As the number of these pulses is equivalent to the unknown measuring voltage, the counter output represents the converted digital word.

In the voltage diagram in Fig. 2.99, T_1 is a fixed time interval, while T_2 is time proportional to the input voltage. The following relationship exists:

$$T_u = n \cdot \frac{1}{f_T}$$

The digital word at the counter output thus represents the ratio of input voltage to reference voltage. The time T_2 is reached in the voltage diagram when the full input voltage is

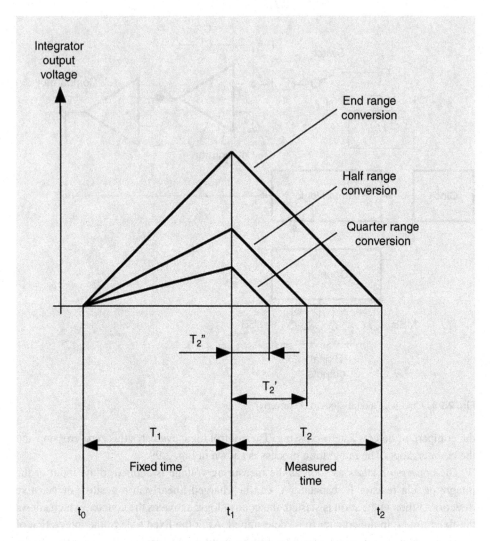

Fig. 2.99 Voltage curve at the output of the integrator of a dual-slope AD converter

applied. The time applies T_2' when half the input voltage is applied, and this is called half range conversion. The duration of the conversion time therefore always depends on the amplitude of the input voltage.

The dual-slope process has several special features: First, the conversion accuracy depends on the stability of the clock generator and the integration capacitor as long as it remains constant during the conversion period. The accuracy depends only on the precision of the reference voltage and the linearity of the integrator. Secondly, if T_1 corresponds to the period of the disturbance, the noise suppression of the converter can be infinite. To suppress a 50 Hz mains hum, T_1 is set to 20 ms.

The achievable conversion rates of a dual-slope converter depend on the clock frequency and the factors determining the integration time, mainly on the capacitance of the integration capacitor.

2.9.6   Voltage to Frequency Converter

Analog-to-digital converters, which operate according to the method of voltage-to-frequency converters, emit at their output a serial pulse train whose frequency is proportional to the analog input voltage. With each voltage change m input of the converter, a frequency change occurs at the output. For the direct representation of the digital value of the analog voltage, additional circuit elements are therefore required, as the circuit of Fig. 2.100 shows.

Due to its conversion method, a VF converter (voltage-frequency converter) has a wide dynamic range of up to seven decades. It can also be deduced from this conversion method that these converters satisfy the conditions of "monotony". Thus, no errors due to missing codes can occur with these converters. Similar to the ramp or slope methods already described, VF converters can suppress noise components in the input signal to a large extent. Common-mode suppression for certain signal frequencies is also possible. A particular advantage of these converters is their ability to transmit digital data directly in series. This is associated with higher immunity to interference on the transmission lines compared to the transmission of analog signals.

In the circuit shown in Fig. 2.101, the mono flop works as a square wave generator, because the output Q is connected to the input. If the mono flop tilts back from its metastable position, a positive signal is generated at the output Q, bringing the mono flop back into the metastable position.

The 74121 module contains a non-retriggerable monovibrator with complementary outputs. The duration of the emitted pulse depends on the time constant $R \cdot C$. The resistor R can range from 2 kΩ to 40 kΩ and the capacitor C from 10 pF to 1000 µF. For lower accuracy requirements it is possible to work without external time components. Only the

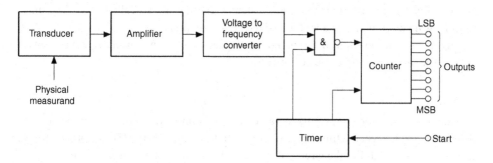

Fig. 2.100 Complete data acquisition system with a VF converter

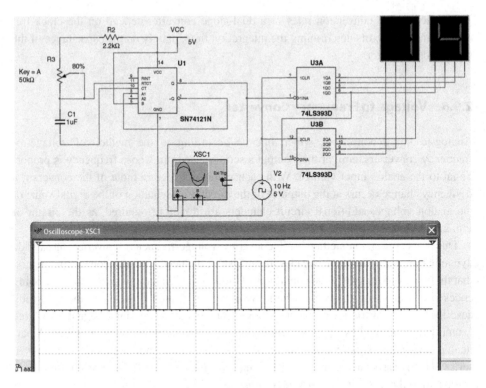

Fig. 2.101   Realization of a VF converter using a mono flop

internal resistor from 2 kΩ is used (pins 9 and 10 connected together, 10 and 11 open), resulting in an output pulse of approx. 30 ns. The pulse duration is largely independent of operating voltage and temperature and is essentially determined by the quality of the time components.

At inputs A1 and A2, the mono flop is triggered with the negative edge of the input signal. The respective other A input and the B input are connected to the ground. Input B has a Schmitt trigger function for slow input edges up to 1 V/s and triggers the monoflop at the 01 transition (positive edge), whereby A1 or A2 must be connected to the ground.

The circuit is not re-triggerable. For a new triggering the so-called "recovery time" must be waited for, which is about 5% of the pulse duration.

The metastable position of the monoflop depends on the two external components R and C. The metastable time is calculated from

$$t_m = 0.7 \cdot R \cdot C$$

Since the capacitor C has a fixed value, e.g. the ohmic value of the photoresistor R determines the metastable time. Instead of the photoresistor, NTC, PTC, or other variable resistors can be used. For the circuit of Fig. 2.100, two monostable times result

$$t_{m_{min}} = 0.7 \cdot 2.2k\Omega \cdot 100nF = 154\,\mu s \Rightarrow 650Hz$$
$$t_{m_{max}} = 0.7 \cdot 52.2k\Omega \cdot 100nF = 3.65ms \Rightarrow 27Hz$$

At a monostable time of $t_{m_{min}} = 154\,\mu s$, 650 pulses per second are generated and the two counters can only store 255 pulses. Therefore, the reset counter is at the frequency of $f = 100$ Hz and the measuring error is small, because only 65 pulses can be counted. At $t_{m_{max}} = 3.65ms$, only 27 pulses are counted. You have to adjust the circuit according to your specifications.

If you have a reset frequency of 10 Hz with a duty cycle of 1:99, all 100 ms will result in a reset pulse for the two 74393 devices.

The voltage-to-frequency converter provides a sequence of serial digital output pulses at the output, as shown in Fig. 2.102. The output frequency is proportional to the change of the photo-resistance. In contrast to conventionally constructed A/D converters, which have a staircase transmission characteristic, the input voltage and output frequency of a voltage-to-frequency converter are in a linear relationship. The transfer characteristic curve, therefore, shows a straight line.

In the realization of a VF converter, a current or voltage source controlled by the unknown input voltage drives the charging current of a capacitor. After reaching a certain capacitor charge, the comparator responds and triggers the downstream monoflop. The latter is required to generate constant pulse widths at output f_a.

Many of the parameter definitions of AD and DA converters also apply to voltage-to-frequency converters. These are Non-linearity, gain error, offset error, and the temperature coefficients of gain offset and linearity. The following parameters describe the special properties of VF converters:

- "Frequency range": The frequency range of a VF converter can be determined by the charge or discharge behavior of the internal integrator. The maximum output frequency (full-scale output frequency) is proportional to the maximum amplitude of the input signal. Furthermore, the higher the maximum possible output frequency, the greater the dynamic range. Usual values of the output frequency are between kHz and MHz.
- "Dynamic range": The logarithmic ratio of the maximum possible operating signals to the minimum possible operating signals is considered by the dynamic range. This is expressed in dB and can be determined both with voltages and with currents or frequencies.
- "Response time": This parameter specifies the maximum time required for the output frequency to stabilize after an amplitude jump is applied to the input. In applications where a high degree of accuracy is required for rapidly changing input signals, this parameter is very important.

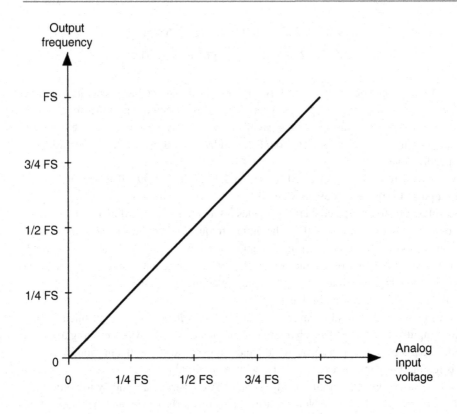

Fig. 2.102 Ideal transfer characteristic of a voltage-frequency converter

Temperature Sensors

3

Summary

Heat and temperature are often confused in everyday language. These are quite different sizes. The atoms and/or molecules that make up every substance are in constant motion, invisible to the human eye. Their speed increases with rising temperature. The temperature thus describes the thermal state.

In contrast, heat is the kinetic energy of molecules and atoms. Energy can be used to do work. This work capacity is called its heat quantity Q; it has the unit "Joule" (J). This is defined as

$$J = N \cdot m = W \cdot s = \frac{kg \cdot m^2}{s^2}.$$

Every substance can assume three different states: solid, liquid, or gaseous—the so-called "states of aggregation". Which of these it takes depends on the temperature. At certain temperatures, two or even three states of aggregation can exist side by side. In the case of water, the temperature at which the solid, liquid, and gaseous phases exist together (the so-called triple point temperature) is 0.01 °C.

A solid body has a certain shape and volume. If a force acts on it, an internal force counteracts the change of shape and volume, which is called "cohesive force". A liquid body has only a certain volume; this adapts to any given shape. The surface is always horizontal when at rest. A cohesive force also acts in liquids, but this is much less than in solid bodies.

A gaseous body, on the other hand, has neither a fixed shape nor a certain volume but fills every available space completely. The volume can be changed by external pressure. Between atoms resp. molecules of a gas, no cohesive forces are working, the cohesive effect is canceled. They are in steady motion without rules.

© Springer Fachmedien Wiesbaden GmbH, part of Springer Nature 2022
H. Bernstein, *Measuring Electronics and Sensors*,
https://doi.org/10.1007/978-3-658-35067-3_3

Fig. 3.1 Typical time course of the solidification curve of a material

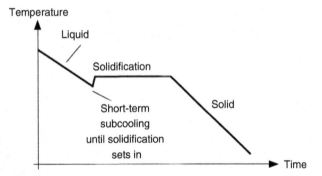

In practice, there are other fixed points besides water. These are the solidification points of pure metals. If a molten metal cools down, the melt begins to solidify at a certain temperature (Fig. 3.1), whereby supercooling can sometimes occur for a short time. The transformation from the liquid to the solid phase is not abrupt but slow. The temperature remains constant until the transformation is completely completed. This temperature is called the solidification temperature, its value depends on the purity of the metal. In this way, certain temperatures can be reproduced with high accuracy.

The temperatures corresponding to the fixed points can be determined with gas thermometers or other measuring instruments. The values are then legally prescribed on the basis of a large number of comparative measurements in the state institutes (e.g. PTB).

3.1   Basic Information About Temperature Measurement

Modern semiconductor components are indispensable for temperature measurement. Today it is possible to integrate all necessary additional functions including amplifier and possibly cold junction on a single silicon chip in addition to the actual sensor element.

There are several possibilities for the output of the measured value, such as in the form of current, voltage, frequency, or as a digital signal, which can then be processed by a connected PC system.

3.1.1   Temperature-Dependent Effects

Almost all physical properties of a substance change with temperature, e.g. The dimensions, density, resistivity, dielectric constant, magnetic susceptibility, sometimes also the color (when considering all wavelengths).

The three most commonly used temperature-dependent electrical quantities are resistance, thermoelectric EMF, and the voltage drop across a current-carrying semiconductor diode. In the USA, the thermocouple is the most common, closely followed by the resistance thermometer. This is the most important sensor for temperature measurement in

Fig. 3.2 Characteristic curves of temperature sensors based on a change in resistance

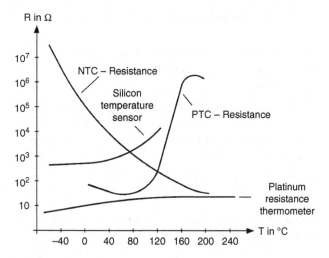

Germany, but thermocouples are also rapidly gaining in importance here. Diodes are used as sensors within temperature measuring integrated circuits. Figure 3.2 summarizes the most important temperature sensors based on a change in resistance.

In practice, the quite different characteristic curves result in different application possibilities. A circuit connected to the sensor evaluates the resistance changes and converts them into a signal form that can be easily processed.

It is important to distinguish between contact and non-contact temperature measurement. Contact measurement is considerably easier and also less expensive because the sensor can be attached directly to the object to be measured. This provides high accuracy, fast response, and a wide temperature range. The non-contact temperature measurement evaluates the infrared radiation emitted by a body. It can be found, for example, at rotary furnaces or blast furnaces in heavy industry and in the chemical industry—wherever it is not possible to measure directly, but only from a greater distance. Suitable sensors are pyrometers, of which there are numerous designs:

- Total radiation pyrometers for wavelength-independent temperature measurement, which work according to the principle of the "Stefan-Boltzmann" radiation law
- Spectral pyrometer for a narrow band range of temperature radiation according to the "Planckian" radiation law
- Strip-beam pyrometers for the broadband range of temperature radiation, which operate according to the "Stefan-Boltzmann" or "Planck" radiation law
- Radiance pyrometers operating according to the comparison method
- Distribution pyrometers which obtain the temperature either from the radiance of a color measurement or from a comparison method in conjunction with a mixed color
- Ratio pyrometers, which determine the measurement result from a series of measurements in different ranges of the spectrum of temperature radiation

3.1.2   Temperature-Dependent Resistors

The resistance value depends on the material, the dimensions, and the temperature. Table 3.1 lists the resistivity ρ and the temperature coefficient α of the most important conductive materials.

The temperature dependence of the ohmic resistors is indicated by the temperature coefficient α. This indicates how much a resistor from 1 Ω increases in temperature at 1 K. The unit of measurement for this is 1/K. Kelvin is the unit of measurement for temperature according to the currently valid system of units. 0 K corresponds to absolute zero, the following temperature conversion applies:

Table 3.1 Specific resistance ρ and temperature coefficients α of various materials. The designation "WM" means resistance material

	Material	ρ in $\dfrac{\Omega \cdot mm^2}{m}$	γ in $\dfrac{m}{\Omega \cdot mm^2}$	α in 1/K
Metals	Aluminium	0.0278	36	0.00403
	Lead	0.2066	4.84	0.0039
	Iron Wire	0.15–0.1	6.7–10	0.0065
	Gold	0.023	43.5	0.0037
	Copper	0.01724	58	0.00393
	Nickel	0.069	14.5	0.006
	Platinum	0.107	9.35	0.0031
	Mercury	0.962	1.04	0.00092
	Silver	0.0164	61	0.0038
	Tantalum	0.1356	7.4	0.0033
	Tungsten	0.055	18.2	0.0044
	Zinc	0.061	16.5	0.0039
	Tin	0.12	8.3	0.0045
Alloys	Wood metal	0.5	2.0	±0.00001
	Constantan (WM 50)	0.43	2.32	0.00001
	Manganese	0.3	3.33	0.00035
	Nickel silver (WM 30)	1.09	0.92	0.00004
	Nickel-chromium	0.43	2.32	0.00023
	Nickeline (WM 43)	0.13	7.7	0.0048
	Steel wire (WM 13)	0.54	1.85	0.0024
Non-metals	Graphite	22	0.046	−0.0002
	Coal	65	0.015	−0.0003
Film resistances	Coal layer to 10 kΩ			−0.0003
	Coal layer to 10 MΩ			−0.002
	Metal layer			±0.00005
	Metal oxide layer			±0.0003

$$0 \text{ K} \triangleq -273.16°\text{C} \quad (\text{Absolute zero-point})$$
$$273.16 \text{ K} \triangleq 0°\text{C}$$
$$293.16 \text{ K} \triangleq 20°\text{C} \quad \text{Zero temperature} \left(\text{Reference temperature}\right)$$

Temperature differences are identical on the Kelvin and Celsius scales. The temperature coefficient depends not only on the material of the resistor but also on the temperature itself, and in the case of carbon film resistors additionally on the resistance value. In the tables it is mostly given for the room temperature 20 °C = 293,16 K.

The resistance value R_T of a conductor at a temperature other than the reference temperature of 20 °C is given by the formulae below:

$$
\begin{aligned}
R_W &= R_{20} + \Delta R & R_W &= \text{Final resistance value in } \Omega \\
&= R_{20} + \left(R_{20} \times \alpha \times \Delta T\right) & R_{20} &= \text{Resistance value at } 20°\text{C in } \Omega \\
&= R_{20} \times \left(1 + \alpha \times \Delta T\right) & \Delta R &= \text{Resistance difference in } \Omega, \\
& & \alpha &= \text{Temperature coeffizient in } 1/\text{K}, \\
& & \Delta T &= \text{Temperature difference in K or }°\text{C}
\end{aligned}
$$

A carbon film resistor with a temperature coefficient of $\alpha = -0.0003/\text{K}$ has a value of 10 kΩ at room temperature. The value at 100 °C is calculated as follows:

$$R_T = R_{20} \times \left(1 + \alpha \times \Delta T\right)$$
$$= 10k\Omega \times \left(1 + \left[-0.0003/\text{K}\right] \times 80\text{K}\right) = 9.760k\Omega$$

The following equations can be used to calculate series and parallel connections of resistors with different temperature coefficients:

Series connection

$$\alpha = \frac{\alpha_1 \cdot R_1 + \alpha_2 \cdot R_2}{R_1 + R_2}$$

Parallel connection

$$\alpha = R \cdot \frac{\alpha_1 \cdot R_2 + \alpha_2 \cdot R_1}{R_1 \cdot R_2}$$

where

$$\alpha = \text{Total temperature coefficient in } 1/\text{K}$$
$$R = \text{Total resistance in } \Omega$$
$$\alpha_1, \alpha_2 = \text{Temperature coefficient of the single resistors in} 1/\text{K}$$
$$R_1, R_2 = \text{Single resistor in } \Omega$$

3.1.3   NTC Resistors or Thermistors

With NTC resistors (negative temperature coefficient) the resistance decreases with increasing temperature, therefore they are also called "hot conductors". The effect is based on the increasing number of free charge carriers with rising temperatures. The temperature coefficient is much higher than with metals—in practice in the order of 3–6% 1/K and therefore these components are often used as temperature sensors.

Figure 3.3 shows the characteristic curve of an NTC resistor with the circuit symbol. The two opposing arrows define the mode of operation of an NTC behavior because the resistance value decreases with increasing temperature. When deciding on a particular type, in almost all cases the design that is optimal for the respective application is taken as a basis. Special hot conductor designs are available for liquids, gases, and solids. In tablet form, they are first completed by the user to form a mountable sensor.

Once the design has been chosen, the nominal resistance value must be determined—the value at nominal temperature, which is usually 25 °C, but sometimes also −30 °C or +100 °C.

The temperature-dependent resistance change cannot be evaluated directly, but only via a measuring voltage. In order to make this as high as possible, high-impedance NTC types are preferred. The higher the voltage at the thermistor and the higher its load capacity, the higher the achievable measurement voltage. The inherent load of the NTC resistor heats it up and thus falsifies the measured value. Therefore, it should not be set too high. The thermal conductivity value G_{th} in mW/K indicates the dead load that heats the NTC resistor around 1 K.

Fig. 3.3 Characteristic curve and switching symbol of an NTC resistor

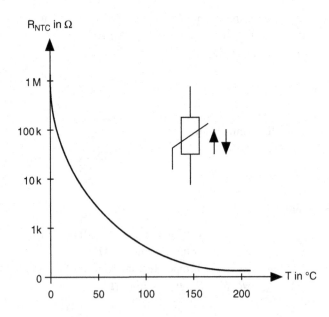

The respective value is always given in the datasheet. Figure 3.4 shows the relationship between temperature and voltage on a type K164 (Siemens) hot conductor at a self-loading of 5 mW. One can see the large differences in the various nominal resistance values. The measurement voltage achievable with a thermistor is of course the higher, the greater the operating voltage applied.

In classical measurement technology, an NTC resistor is operated in a measuring bridge, as the circuit of Fig. 3.5 shows. It is adjusted by the R_4 adjuster at 20 °C. At the

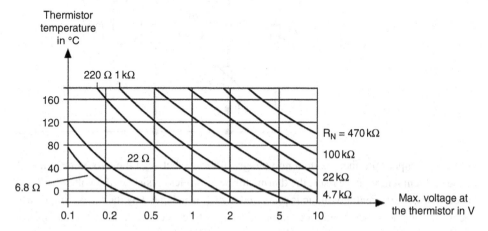

Fig. 3.4 Permissible voltage at hot conductors of type K164 (Siemens) with different nominal resistances RN at constant load with PNTC = 5 mW

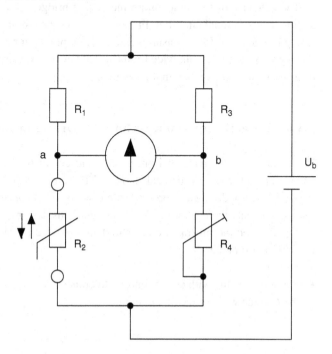

Fig. 3.5 Temperature measuring bridge with NTC resistor

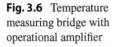

Fig. 3.6 Temperature measuring bridge with operational amplifier

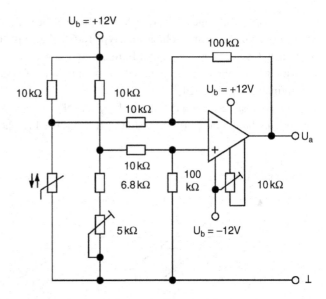

reference temperature there is then no voltage difference between the two points a and b. If the ambient temperature rises, the thermistor becomes less resistive; the voltage at point a decreases. A differential voltage is produced, which is indicated by the measuring instrument. If, on the other hand, the ambient temperature decreases, then the thermistor increases its resistance value, and the voltage at point b rises. The scale in the measuring instrument can be labeled directly in °C, thus providing an analog temperature display.

The behavior of the temperature measuring bridge can be improved considerably by adding an operational amplifier. Figure 3.6 shows the circuit. The left branch of the measuring bridge is used for measurement, the right branch for adjustment. The amplification factor can be varied within wide limits and, if necessary, can be adjusted so that precision measurements can be made in narrow ranges.

3.1.4   Data, Designs, and Technology of Thermistors

Hot conductors are made of iron, nickel, and cobalt oxides, to which other oxides are added to increase the mechanical stability. They are prepared into a powdery mass, mixed with a binder, hydraulically pressed into discs or other forms under a pressure of several tons per cm, and then sintered. There are numerous designs, as shown in Fig. 3.7.

The finished thermistors are measured and selected for close tolerances. Two parameters are important here:

- The $R_{25\,°C}$ -value with corresponding tolerances
- the B-value with its specific tolerance

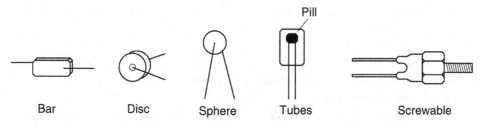

Fig. 3.7 Commercially available types of hot conductors

It is therefore not sufficient to measure only the resistance at 25 °C. The tolerance of the B-value has a much stronger influence on accuracy. It is a measure of the temperature dependency of the thermistor and, according to DIN 44070, is always referred to as two measurement temperatures, namely +25 °C and +85 °C. This results in the steepness of the characteristic curve. The greater the $B_{25/85}$-value, the more sensitive the sensor is, expressed in terms of the negative temperature coefficient

$$\alpha_R = \frac{-B}{T^2} \text{ in } \% / K \qquad \alpha_R = \text{Temperature coeffizient in } 1/K,$$

$$B = \text{Material constant in } K$$

$$R_1 = R_2 \cdot e^{-B\left(\frac{1}{T_1} - \frac{1}{T_2}\right)} \qquad R_1 = \text{Semiconductor resistance at the temperature } T_1 \text{ in } K$$

$$R_1 = R_2 \cdot e^{\alpha_R \cdot \Delta T \cdot \frac{T_2}{T_1}} \qquad R_2 = \text{Semiconductor resistance at reference temperature } T_2 \text{ in } K$$

If one assumes a B-value tolerance of 5%, as is customary with conventional press technology, and selects the sensor at 25 °C to an R-tolerance of ±5% as well, then at 85 °C already results in a total tolerance of ±17%.

In addition, the drift behavior of hot conductors under thermal-cycling conditions causes quite different problems. An example shall clarify which changes can occur here. An already pre-aged standard disc-type hotline (5 mm Ø) with an R25 tolerance and a B-value tolerance of ±5% each was investigated. Under different conditions the following resistance changes occur (R_{25}/R_{85}):

- Dry heat (+125 °C, 1000 h) −2.5%
- Humid heat (+40 °C, 40% relative humidity) −1.5%
- Temperature change (+25 °C/+125 °C, 50 cycles) −0.8%

The characteristic curve shifts significantly due to drift behavior. As a result, the initial tolerance (e.g. ±5%) can deteriorate so much after a relatively short period of operation that the resulting measurement error ΔT is no longer acceptable for many applications, especially in the automotive, household, and industrial electronics sectors.

The following objectives were therefore set in the development of the temperature sensors:

- the optimization of the composition of the ceramic masses, so that the B-value tolerance is reduced to well below 5%, thus significantly increasing the measuring accuracy over the entire temperature range
- better compliance and reproducibility of the nominal characteristic $R = f(T)$ in series production,
- high long-term stability,
- small dimensions for short reaction times.

In order to meet these requirements, the basic prerequisites must be created already during the composition of the ceramic masses. Only very specific compounds can therefore be used for production. These include purest metal oxides or oxidic mixed crystals with a common oxygen lattice. In some cases, stabilizing iron oxides are added, which are relatively insensitive to temperature fluctuations during the production process (sintering). With this, easily reproducible and stable electrical data can be achieved, so that almost identical resistance or temperature characteristics are obtained for each hot conductor with e.g. a certain R25 value, even with large quantities.

3.1.5   Linearization of Thermistor Curves

Thermistors have a strongly non-linear resistance curve. If the temperature measurement in a certain range requires a preferably linear response, e.g. for a scale, the linearity can be noticeably improved with a series and a parallel resistor. Depending on requirements, the temperature range to be measured should not exceed 50–100 K.

A useful measurement voltage curve is obtained if the series resistance R_1 in the circuit shown in Fig. 3.8 is exactly as high as the hot plate resistance R_T in the middle of the temperature range to be linearized. The resistance R_2 parallel to the thermistor must be 10

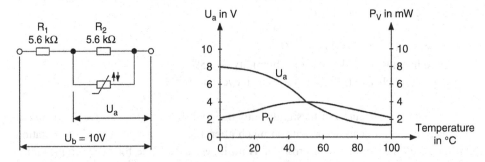

Fig. 3.8 Linearization of a thermistor by an additional resistor circuit. The diagram shows the linearized curve of the voltage on the hot conductor and the power loss

times as high. Figure 3.8 also contains the diagram for resistance compensation for the thermistor. The general formula for any given thermocouple and temperature range is

$$U_a = \frac{U_b}{\left[R_1 \cdot \left(\dfrac{1}{R_T} + \dfrac{1}{R_2} \right) + 1 \right]}$$

$$B = \left(\frac{1}{T} - \frac{1}{T_N} \right)$$

where

$$R_T = R_N \cdot e$$

R_N, B, T_N are data sheet values of the hot conductor

U_b is the driving voltage

U_a is the voltage at the temperature T

T is the temperature in $°C$

The sensitivity of the thermistor naturally decreases with linearization.

3.1.6 Amplifying Circuits for Linearized Thermistors

The linearized measuring voltage of the circuit of Fig. 3.8 must not be loaded or only very slightly loaded. Operational amplifiers are used for amplification. Figure 3.9 shows the typical basic circuit and the course of the different resistance ratios R_2/R_1. The amplifier formula for U_a is

$$U_a = U_e \left(1 + \frac{R_2}{R_1} \right) - U_0 \cdot \frac{R_2}{R_1}.$$

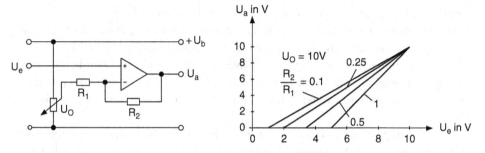

Fig. 3.9 Basic circuit for amplifying the hot wire voltage. The diagram shows the influence of the resistance ratio R_2/R_1 on the output voltage

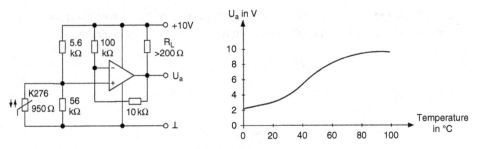

Fig. 3.10 Linearization of the hot conductor characteristic by means of a simple amplifier circuit. The diagram shows the output voltage at the load resistor as a function of temperature

Fig. 3.11 Threshold amplifier with a switching threshold at 80 °C

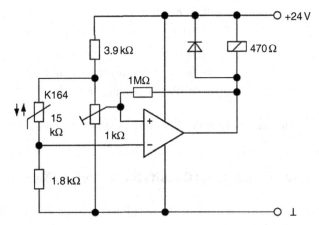

In practice, the circuit shown in Fig. 3.10 is used to control a relay. The operational amplifier operates in non-inverting mode, the gain is adjusted as shown in Fig. 3.9. Since the operational amplifier has an "open collector" as output, the working resistance can be driven directly. If a relay is used, a diode must be switched on in parallel so that the transistor in the OP output stage is not damaged by voltage peaks from the relay coil.

The remaining non-linearity is also transferred to the load resistor. If you want to have a decreasing voltage on the load resistor as the temperature rises, you must connect the thermistor to the positive supply voltage and the load resistor to the ground.

In many thermistor applications, only a certain temperature threshold is to be detected, e.g. if a fan is to switch on for cooling at a high temperature. The optimum switching sensitivity is achieved when the highest possible voltage is applied to the thermistor. As already mentioned, the highest thermistor voltage is limited by the maximum power dissipation. For this reason, a voltage divider is usually used, as in the threshold amplifier in Fig. 3.11.

The operating voltage in this circuit is 24 V, the permissible voltage at the thermistor should not exceed 2 V. Threshold value switches always require feedback to the non-inverting input so that no unwanted oscillation occurs near the switching point. A resistor from 1 MΩ is used here for this purpose. This automatically creates a switching hysteresis.

3.1.7 PTC Resistors

PTC resistors have a positive temperature coefficient, i.e. the resistance value increases with rising temperature. For this reason, the circuit symbol has two arrows pointing in the same direction. The PTC resistors are also called "PTC thermistors". Almost all metals are PTC thermistors; the resistance changes caused by temperature fluctuations are very small here. They are much higher in certain sintered ceramics, such as barium titanate and similar titanium compounds. The electrical properties can be varied by adding metal oxides and salts.

In Fig. 3.12 the very steep rise in the resistance/temperature characteristic curve illustrates the particular advantage of the ceramic PTC thermistor as a temperature sensor, e.g. in the monitoring of fixed limit temperatures. Here it can be used for measurement and control tasks with little effort. The PTC thermistor is in thermal contact with the body or medium to be monitored. If the specified limit temperature is exceeded, it becomes abruptly highly resistive—with an increase in resistance of up to several powers of ten.

As can be seen from a closer look at the characteristic curve of a PTC thermistor, there is a thermistor behavior up to the initial temperature T_A. After that, the temperature

Fig. 3.12 Characteristic curve and symbol of a PTC resistor

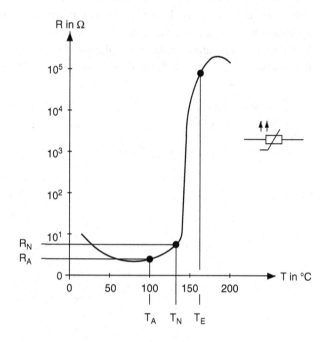

coefficient changes its sign. At the nominal temperature T_N, the steep rise of the curve begins. The nominal resistance results as follows:

$$R_N = 2 \times R_A$$

At the final temperature T_E, the steep rise of the curve ends. Above the final temperature, the curve becomes flatter and changes back to a thermistor behavior.

3.1.8 Protective Circuits with PTC Thermistors

With PTC thermistors, electrical machines can be protected very reliably and effectively against thermal overload. An overcurrent occurring in the event of a fault or an exceeding of the maximum permissible temperature does not lead to destruction.

The motor protection sensors shown in Fig. 3.13 are installed directly in the winding so that the good heat coupling enables a fast and reliable evaluation of a motor malfunction. The response temperature is selected so that the PTC thermistor suddenly becomes highly resistive if the maximum permissible operating temperature of the motor is exceeded. Three PTC sensors are connected in series in three-phase motors. A connected evaluation circuit separates the motor from the mains in the event of thermal overload by switching off the motor contactor. By design measures, a high response sensitivity is achieved, which allows a simple evaluation circuit.

If a PTC thermistor is connected in series with a load to be protected, the power loss generated in it is directly dependent on its current consumption. It is dimensioned so that it does not heat up noticeably at rated current. It, therefore, has a low resistance at nominal operation and only a low voltage drops across it. In the event of a fault, it heats up above its reference temperature and becomes abruptly highly resistive. This considerably reduces the current flowing through the load, and almost the entire operating voltage of the circuit

Fig. 3.13 Machine protection sensor for three-phase motors with PTC resistor

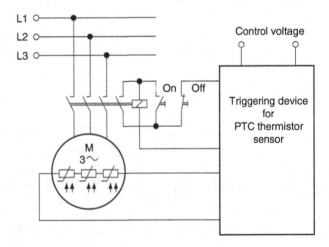

Fig. 3.14 Overcurrent protection using PTC thermistors

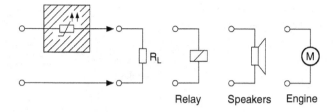

Relay   Speakers   Engine

now drops at the PTC thermistor. The low residual current does not load the device to be protected but is sufficient to keep the PTC thermistor high-resistant. This state is maintained until the voltage is switched off and the PTC thermistor cools down.

Figure 3.14 gives simple, effective options for automatic short-circuit protection or overcurrent protection with a PTC resistor. The possible causes for the PTC thermistor to respond as a protective element in the event of a fault can be not only overcurrent or over-temperature, but also combinations of both. The PTC thermistor is therefore temperature and current sensitive protective component that effectively and safely prevents an overload of an electrical or electronic device.

3.1.9 Temperature Switches from −10 °C to +100 °C

In practice, the silicon temperature sensor KTY10 is used. The KTY10 is suitable for temperature measurement in gases and liquids in the range from −50 °C to +150 °C. Figure 3.15 shows the connection diagram of the temperature sensor.

The sensor element consists of N-conducting silicon in planar technology. The slightly curved characteristic $R_T = f(T_A)$ of Fig. 3.16 is described via the regression parameters. The resistance can thus be calculated for different temperatures according to the following second-degree equation in the temperature range from −30 °C to +130 °C:

$$R_T = R_{25} \times \left(1 + \alpha \times \Delta T + \beta \times \Delta T_A^2\right) = f\left(T_A\right)$$

$\alpha = 7.68 \times 10^{-3}$ and $\beta = 1.88 \times 10^{-5}$ (typical curve from Fig. 3.16).

The temperature factor k_T can then be determined from this.

In order to protect the connections from moisture, e.g. when measuring liquids, they should be covered with shrink tubing or a hot-melt adhesive. In addition, care must be taken to ensure that the sensor has good thermal contact with the monitored parts.

The simple characteristic linearization with an optimal resistance leads to linearity errors below 0.6 K in the range from −40 to +130 °C. In order to fully exploit the advantage of the small chip dimensions with respect to the thermal time constant of the sensor, application-specific package designs are required. This enables a broad application of the new sensor in the field of measurement and control technology, automotive engineering, and the consumer appliance market.

Fig. 3.15 Connection diagram
of the temperature sensor

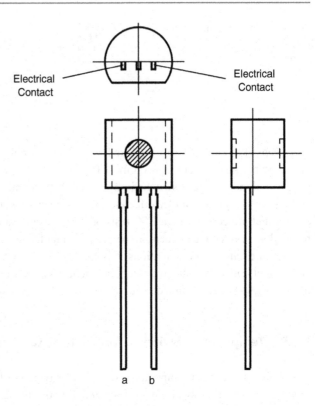

Fig. 3.16 Characteristic curve
of the silicon temperature
sensor KTY10

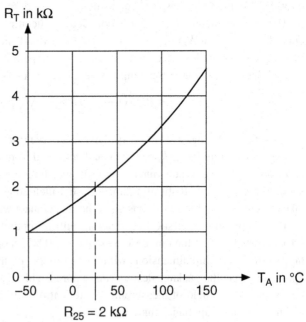

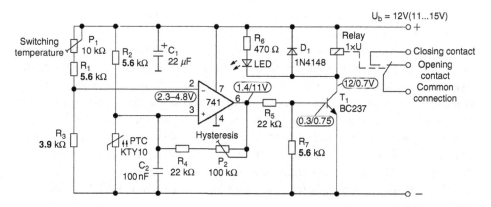

Fig. 3.17 Temperature switch from −10 to +100 °C, whereby the switching temperature and hysteresis can be continuously adjusted

The temperature switch shown in Fig. 3.17 switches on or off at a temperature pre-selected by a potentiometer. The switching threshold can be continuously adjusted in a range from approx. −10 to +100 °C. In addition, the switching hysteresis can be changed via a potentiometer. If the temperature exceeds or falls below the set value, the relevant device is switched on or off.

In the circuit shown in Fig. 3.17, the temperature is measured with the silicon temperature sensor KTY10. This component, which looks like a normal transistor, has a positive temperature coefficient, so it behaves like a PTC resistor. Together with resistor R_2, it forms a voltage divider which emits a voltage dependent on the actual temperature. This voltage is compared in the operational amplifier 741, which is connected as a comparator, with the nominal temperature voltage from R_1, R_3, and P_1. The output of the operational amplifier (pin 6) remains at the "0" signal as long as the actual voltage at the non-inverting input (pin 2) is lower than the set voltage at the inverting input (pin 3). If the voltage at the inverting input exceeds (due to temperature rise) the set voltage (set temperature) set at the non-inverting input, the output switches to the "1" signal. At the output of the operational amplifier, there is a transistor that amplifies the low output current of the operational amplifier. The "1" signal makes it conductive, the relay picks up and the LED emits light. At the same time, the voltage is coupled via resistor R_4 and P_2 in order to create a certain switching hysteresis and thus prevent the relay from fluttering.

With the adjuster P_2, the switching hysteresis (temperature difference between "On" and "Off") can be influenced. A reduction in P_2 increases hysteresis. If the temperature rises, the relay picks up, if it drops, it drops out.

Before the circuit is put into operation, the following settings must be made: First, the slider of adjuster P_1 is moved to approximately the middle position. The slider of adjuster P_2 must be turned to the lowest resistance value.

A DC voltage source from +12 V is required to operate the temperature switch, in practice a ready-made power supply unit with a fixed voltage regulator. For safety reasons, the power supply unit must comply with VDE regulations. It is important to observe the polarity, as any mix-up will inevitably destroy the circuit.

After connecting the operating voltage, turn the slider of adjuster P_1 to the left and to the right until the stop. The relay must pick up and release alternately. The LED must signal the switching process in the same alternation. If not, the operating voltage must be switched off immediately and the circuit must be checked again. If the circuit works according to these conditions, check whether the following voltages are applied to the pins of the integrated circuit: Pin 2: 2.3–4.8 V (this voltage depends on the position of the adjuster P_1; it should be possible to adjust it within the indicated range by adjusting P_1); Pin 3: approx. 3.0 V (at a room temperature of approx. 20 °C); Pin 6: With the relay not energized approx. 1.4 V; with the relay energized approx. 11 V. Pin 4: 0 V; Pin 7 = Operating voltage +12 V.

Since the contacts of the relay are potential-free, the temperature switch can be used universally. If change-over contacts are selected here, devices can be switched on or off as required when the temperature is exceeded or undercut. If you want to monitor a freezing point, insert the sensor in ice water that you have previously measured with a thermometer and set the switching temperature with the adjuster P_1. If the circuit is to be adjusted to room temperature, place the sensor in lukewarm water at 25 °C. In this way, almost any switching threshold can be set. In any case, you should wait a few minutes until the sensor has reached exactly its ambient temperature.

3.1.10 Temperature Switch with Sensor Monitoring

Many temperature switches have a disadvantage: If the sensor cable is interrupted or short-circuited by a defect, the heating or cooling remains switched on, which can lead to overheating or undercooling and thus to considerable damage. The temperature circuit with integrated sensor monitoring (short-circuit/interruption) presented here switches on a relay, which then switches on a heater fan, frost monitor, or similar devices when the temperature falls below the set temperature. The switching temperature is adjustable, four LEDs signal the individual switching states.

Automatic temperature switches are usually designed as so-called two-point controllers. This is a circuit that simply compares a given setpoint (the desired temperature) with the actual prevailing actual value; if the actual value is below the setpoint, heating is switched on (this is called operating point AP1). As soon as the setpoint is reached, the heating is switched off again (operating point AP2). The other case, that it is warmer than desired, is usually not taken into account, as this hardly occurs in our latitudes anyway. If this should also be ensured, then a cooling device should be switched on at such moments.

Figure 3.18 shows the mode of operation of a two-point controller. Important parameters here are the time constant T_s and the dead time T_t. If the controlled variable x exceeds the reference variable w, the manipulated variable y remains switched on up to operating point A ($w + x_0$) due to the hysteresis of the controller. The controlled variable increases

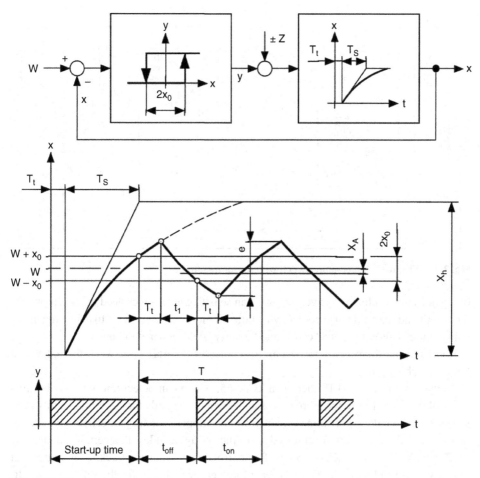

Fig. 3.18 Diagram of a two-point control

further, however, and only decreases after the dead time T_t has elapsed according to the time constant T. The controlled variable is switched on again when the value $w - x_0$ is undershot so that the controlled variable continuously moves around the setpoint with the period T and amplitude x_0 in the steady-state. The mean value of this movement deviates from the setpoint w by the P-deviation X_A, and the difference "e" defines the hysteresis.

Figuratively speaking, the two-point controller therefore constantly moves around the setpoint w. It only knows the states greater or smaller and reacts accordingly only by switching the circuit on and off. This has two principle-related disadvantages: As soon as the larger/smaller detection is only sensitive enough, even the brief response of the heater is sufficient to detect slight warming and immediately switch it off again. A constant rattling of the relay in question would be the result.

Therefore, a dead band between the switching points is introduced, the hysteresis, but this results in a second disadvantage: The heater should work for a while before it switches

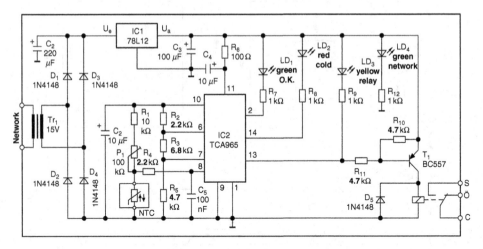

Fig. 3.19 Electronic temperature switch with sensor monitoring

off again. This results in stronger heating than sensitive electronics need to switch off. On the other hand, it should also cool down a little bit more than necessary to detect the under temperature. This results in a "calm" and "orderly" control behavior, but at the expense of a dead band, i.e. there is a temperature difference of several °C between the upper and lower switching point.

In the circuit of Fig. 3.19 there is a TCA965, which can detect not only one but two thresholds. In contrast to the more or less undefined dead band of a hysteresis, in this case, we speak of a window discriminator. The lower and upper switching thresholds, which can be set in a defined manner, form the edges of the voltage window to a certain extent.

The TCA965 is particularly suitable for control engineering as a tracking or adjustment circuit with a dead band, as well as in measurement technology for the selection of voltages that should be within a certain tolerance range from the required setpoint.

The window discriminator of Fig. 3.20 analyzes the magnitude of the input voltage with reference to two limits, which are entered as voltages from outside. The window within which the circuit reacts with "good" can be entered either by an upper (U_6) and a lower limit (U_7) or by the center of the window (U_8) and, depending on this, by a voltage ΔU (U_8), which corresponds to half the window width and is entered against the ground. The changeover points have a Schmitt trigger characteristic with a small hysteresis.

The module provides four output signals: Input signal inside or outside the window (good, bad) or too high or too low. All outputs have open collectors which can be loaded with up to 50 mA. This allows direct control of light-emitting diodes or small relays. Furthermore, the TCA965 also contains a reference voltage from which all voltage thresholds can be derived—largely independent of temperature and operating voltage.

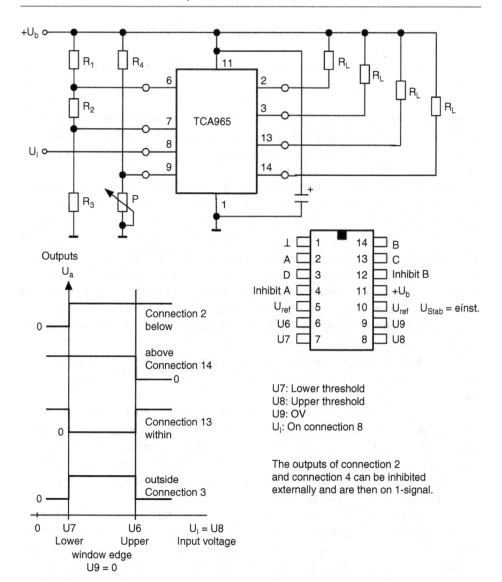

Fig. 3.20 Connection diagram, basic circuit and logical behavior of the TCA965

In the TCA965 of Fig. 3.20, the upper threshold (U_{max} or U_{min}) is located at pin 6 and the lower threshold (U_{max} or U_{min}) at pin 7; the voltage to be monitored, U_a, goes to pin 8. The device detects the following states:

1. U_a is below U_{min}, output A or pin 2 active   **U_{in}**
2. U_a is above U_{max}, output B or pin 14 active

3. U_a is above U_{min} and below U_{max}, i.e. within the specified window, output C or pin 13 active
4. U_a is below U_{min} or above U_{max}, i.e. outside the specified window, output D or pin 3 active.

All four outputs are 0-active, i.e. they switch through a transistor with an open collector. By the way, state 4 defines the inverting connection of state 3, because only either one or the other can be "true". For this reason, no further evaluation of the signal from output D (pin 3) was used in the circuit of Fig. 3.20.

The window discriminator TCA965 itself is decoupled from the rest of the circuit of Fig. 3.19 again via the RC element R_6/C_4. Possible feedback effects due to output-side changeover processes at the relay or at the LEDs thus remain without effect on the internal evaluation logic. At pin 10 the TCA965 generates a further decoupled, internally stabilized voltage of +6 V. This voltage serves as a constant reference voltage for the two dividers R_2/R_3 and R_1/P_1/NTC and ensures that absolutely no unwanted interference can be interfered with. If a clear behavior is required at the moment of switching of the TCA965, and no significant hysteresis is to disturb the operating mode, then the threshold values used for comparison must not change in the slightest. This is ensured by the mentioned external and internal circuit measures.

The backup capacitor C_2 and the ceramic capacitor C_5 contribute to these measures. Capacitor C_5 ensures that any interference peaks from the external NTC supply line are short-circuited to ground.

The +6 V provided by the IC output pin 10 is divided by the series connection of R_2, R_3 and R_5 in such a way that pin 6 is connected to approx. 5 V, which corresponds to the upper window edge, and pin 7 to approx. 2 V, which corresponds to the lower window edge. So if the input voltage at pin 8 is below 2 V because the thermistor is correspondingly warm and thus has low resistance, then output A (pin 2) is active. The green LED 1 lights up and signals that the set temperature has been reached.

Assuming that this is to be the case at 20 °C, where the NTC resistor has just reached its nominal value of 25 kΩ, then R_1 and P_1 together would have to be twice as large as the NTC resistor to divide 6 V from pin 10 to 2 V for pin 8. This is the case with the adjuster P_1 (100 kΩ) when 40 kΩ is set, which is approximately the middle position, i.e. 40 kΩ of P_1 plus 10 kΩ of R_1 give twice the 25 kΩ of the NTC resistor at 20 °C. An exact scaling of the adjuster is of course not possible due to the component tolerances, but this rough calculation provides an approximate guide for the basic setting.

If the input voltage at pin 8 rises to 2 V, which depends on the ambient temperature and the P_1 setting, then output 13 and LED 3 are active, i.e. heating must be restored and the relay picks up. This is the case when the resistance of the NTC increases due to cooling. Note the opposite trend.

If it gets so cold at the NTC resistor that there is more than 5 V at pin 8, the upper threshold is exceeded and output B (pin 14) switches through. The red LED emits light and signals that it is too cold. This can happen, for example, if the heater is defective or if there is insufficient heat supply. The user can also visually detect this condition by the red LED.

Together with the switching through of output C (pin 13) when the lower voltage source is exceeded or the set temperature has just fallen below the set temperature, the relay is activated. If output C has a 0 signal, i.e. the internal transistor switches through, the BC557 also switches through.

The potential-free changeover contact of the relay can now switch on the heating via the make contact between terminals C and S. There are two things to be considered: First, these contacts may be loaded with a maximum of 500 VA; the voltage should be at a maximum of 250 V and the current should not exceed the limit of 8 A, but not both at the same time. With the mains voltage of 230/240 V, the maximum current flow is 2 A.

Secondly, you must not simply remove the mains voltage from the transformer in order to feed it to a mains-operated heating system; to do this, the mains supply line would have to have a larger cross-section than the small 1 VA transformer requires. Furthermore, a perfect connection with the green-yellow protective conductor must be ensured. Since the switching capacity of 500 VA is not sufficient for larger heaters anyway, it is recommended to use an additional power relay for mains operation.

A function test can be performed as follows: If there is no NTC resistor (open socket BU1), the circuit assumes a very low temperature; then the red LED must light up in any case. If the input socket is short-circuited, the evaluation logic assumes a very low value of the NTC resistor, i.e. the temperature is reached or exceeded. In this case, the green LED LED 1 must light up. The circuit only reacts to an adjustment of the adjuster when the intended thermistor is connected with a supply line up to approx. 10 m. Turning it to the left will only switch on the heater at low temperatures, turning it clockwise will shift the switching point to a higher temperature.

3.2   LED Thermometer

To display a temperature measured with the KTY10 sensor, the circuit of Fig. 3.21 is suitable. This electronic thermometer uses 13 mm high red seven-segment LED displays, it can be used wherever temperatures from −50 to +150 °C have to be measured with great accuracy. The heart of the circuit is the ICL7107, which is largely identical to the ICL7106 except for the display control, is described in detail as it is a standard device in practice. The seven-segment displays must have common anodes, otherwise, the circuit will not work. The thermometer can be adjusted with high precision using two-spindle adjusters. The circuit can be used for measuring the room and outdoor temperature, for heating supply/return as well as in cars, boats, mobile homes, weekend houses, laboratories, air conditioning, industry, trade, etc.

In some applications, especially where the AD converter is connected to a sensor, a different scaling factor between the input voltage and the digital display is required. In a weighing system e.g. the developer may want a full scale when the input voltage has

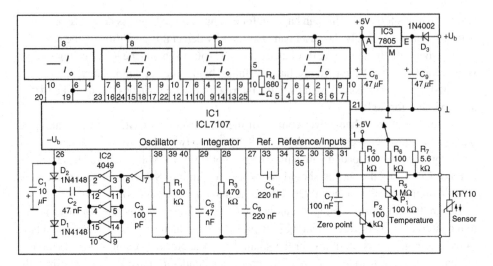

Fig. 3.21 Electronic thermometer with the ICL7107 and four LED-7 segment displays

reached a value of $U_e = 0.682$ V. Instead of a prescaler, which divides the input down to 200 mV, in this case, it is better to use a reference voltage of 0.381 V. Suitable values for the integration elements (resistor and capacitor) in this case are $R = 120$ kΩ and $C = 220$ nF. These values make the system a bit quieter and avoid a divider network at the input. Another advantage of this system is that in one case a "zero indication" is possible at any value of the input voltage. Temperature measuring and weighing systems are examples of this. This "offset" in the display can be easily created by connecting the sensor between "IN HI" and "COM" and by placing the variable or constant operating voltage between "COM" and "IN LO".

3.2.1   Integrated Converter ICL7106 and ICL7107

The circuit ICL7106 and ICL7107 (formerly Intersil, now Maxim) is a monolithic CMOS-AD converter of the integrating type, in which all necessary active elements such as BCD-7 segment decoder, driver stages for the display, reference voltage, and complete clock generation are implemented on the chip. The ICL7106 is designed for operation with a liquid crystal display. The ICL7107 is largely identical to the ICL7106 and directly drives seven-segment LED displays.

ICL7106 and ICL7107 are a good combination of high accuracy, universal application, and economy. The high accuracy is achieved by using an automatic zero balance to less than 10 µV, realizing a zero-drift of less than 1 µV per °C, reducing the input current to 10 pA, and limiting the "roll-over" error to less than one digit.

The differential amplifier inputs and the reference as well as the input allow an extremely flexible realization of a measuring system. They give the user the possibility of bridge measurements, as it is usual for example, when using strain gauges and similar sensor elements. Externally, only a few passive elements, the display, and an operating voltage are required to realize a complete 3 1/2-digit digital voltmeter.

Each measurement cycle on the ICL7106 and ICL7107 is divided into three phases and these are:

- Automatic zero adjustment
- Signal Integration
- Reference integration or deintegration

3.2.1.1 Automatic Zero Adjustment

The differential inputs of the signal input are internally separated from the connections by analog switches and shorted with "ANALOG COMMON". The reference capacitor is charged to the reference voltage. A feedback loop between the comparator output and inverting input of the integrator is closed to charge the "AUTO-ZERO" capacitor $C_{AZ\,in}$ such a way that the offset voltages from the input amplifier, integrator, and comparator are compensated. Since the comparator is also included in this feedback loop, the accuracy of the automatic zeroing is only limited by the noise of the system. In any case, the offset voltage related to the input is lower than 10 µV. Figure 3.22 shows the circuit for the analog section in ICL7106 and ICL7107.

3.2.1.2 Signal Integration

During the signal integration phase, the zeroing feedback is opened, the internal short circuits are removed and the input is connected to the external terminals. The system then integrates the differential input voltage between "INPUT HIGH" and "INPUT LOW" for

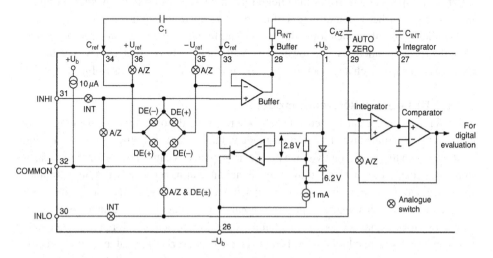

Fig. 3.22 Analog section of ICL7106 and ICL7107

a fixed time interval. This differential input voltage can be within the entire common-mode voltage range of the system. If, on the other hand, the input signal has no reference relative to the power supply, the "INPUT LOW" line can be connected to "ANALOG COMMON" to set the correct common-mode voltage. At the end of the signal integration phase, the polarity of the input signal is determined.

3.2.1.3 Reference Integration or Deintegration

The last phase of the measurement cycle is the reference integration or deintegration. "INPUT LOW" is internally connected to "ANALOG COMMON" by analog switches and "INPUT HIGH" is connected to the C_{ref} reference capacitor charged in the "AUTO-ZERO" phase. An internal logic ensures that this capacitor is connected to the input with the correct polarity, i.e. it is determined by the polarity of the input signal to carry out the deintegration towards "0 V". The time required for the integrator output to return to "0 V" is proportional to the magnitude of the input signal. The digital representation has been specially chosen for 1000 (U_{in}/U_{ref}).

3.2.1.4 Differential Input

Differential voltages can be applied to the input, which are somewhere within the common-mode voltage range of the input amplifier. However, the voltage ranges are better in the range between positive supply from −0.5 V to negative supply from +1 V. In this range the system has a common-mode voltage suppression of typically 86 dB.

However, since the integrator output also oscillates within the common-mode voltage range, it must be ensured that the integrator output does not enter the saturation range. The worst case is when a large positive common-mode voltage combined with a negative differential input voltage in the range of the final value is applied to the input. The negative differential input voltage drives the integrator output in addition to the positive common-mode voltage further in the direction of the positive operating voltage.

For these critical applications, the output amplitude of the integrator can be reduced from the recommended 2 V to a lower value without a great loss of accuracy. The integrator output can approach any operating voltage up to 0.3 V without loss of linearity.

3.2.1.5 Differential Reference Input

The reference voltage can be generated anywhere within the operating voltage range of the converter. The main cause of a common-mode voltage error is a "roll-over error" (different readings when the same input voltage is reversed), which is caused by the reference capacitor being charged or discharged by stray capacitance at its terminals. If a high common-mode voltage is present, the reference capacitor can be charged (the voltage increases) when connected to disintegrate a positive signal. On the other hand, it can be discharged if a negative input signal is to be disintegrated. This different behavior for positive and negative input voltages results in a "roll-over" error. However, if the value of the reference capacitance is chosen large enough, this error can be reduced to less than half a digit.

3.2.1.6 "ANALOG COMMON"

This connection is primarily intended to determine the common-mode voltage for battery operation (7106) or for a system with "floating" inputs relative to the operating voltage. The value is typically about 2.8 V below the positive operating voltage. This value is therefore selected to ensure that 6 V is supplied with power when the battery is discharged. Furthermore, this connection has a certain similarity to a reference voltage. If the operating voltage is high enough to take advantage of the control characteristics of the internal Z-diode ($\approx$7 V), the voltage at the "ANALOG COMMON" connection has a low voltage coefficient. To achieve optimum operating conditions, the external Z-diode with a low impedance (approx. 15 Ω) should have a temperature coefficient of less than 80 ppm/°C.

On the other hand, the limitations of this "integrated reference" should be recognized. With the ICL7107 type, internal heating by the currents of the LED drivers can worsen the characteristics. Due to the higher thermal resistance, plastic encapsulated circuits are less favorable in this respect than those in a ceramic package. If an external reference is used, the ICL7107 does not have any problems either. The voltage at "ANALOG COMMON" is the voltage applied to the input during the phase of automatic zeroing and deintegration. If the "INPUT LOW" terminal is connected to a voltage other than "ANALOG COMMON", the result is a common mode voltage in the system, which is compensated by the system's excellent common-mode voltage suppression.

In some applications, the "INPUT LOW" connector will be set to a fixed voltage (e.g. Reference of operating voltages). It is advisable to connect the "ANALOG COMMON" connector to the same point to eliminate the common-mode voltage for the converter. The same applies to the reference voltage. If it is easy to apply the reference with reference to "ANALOG COMMON", it should be done to eliminate common-mode voltages for the reference system.

Within the circuit, the "ANALOG COMMON" terminal is connected to an N-channel field-effect transistor, which is capable of keeping the 2.8 V terminal below the operating voltage even with input currents from 30 mA or more (if e.g. a load tries to "pull up" this terminal). On the other hand, this terminal only supplies 10 µA as output current, so it can easily be connected to a negative voltage to switch off the internal reference.

3.2.1.7 Test

The "TEST" connection has two functions. In the ICL7106 it is connected to the internally generated digital operating voltage via a resistor from 500 Ω (470 Ω). Thus it can be used as a negative operating voltage for external additional segment drivers (decimal points etc.).

The second function is that of a "lamp test". If this connection is connected to the positive operating voltage, all segments are switched on and the display shows −1888. Caution: In this operating mode, a DC voltage (not a square-wave voltage) is applied to the segments of the 7106. If the circuit is operated in this operating mode for a few minutes, the display may be destroyed!

In 7106, the internal reference of the digital operating voltage is formed by a Z-diode with 6 V and a P-channel "SOURCE follower" of large geometry. This supply is designed to be stable to be able to deliver the relatively large capacitive currents that occur when the backplane of the LCD display is switched.

The frequency of the square wave, with which the backplane of the display is switched, is generated from the clock frequency by dividing it by a factor of 800. At a recommended external clock frequency of 50 kHz, this signal has a frequency of 62.5 Hz with a nominal amplitude of 5 V. The segments are driven with the same frequency and amplitude and are in phase with the BP (backplane) signal when the segments are switched off, or in anti-phase when the segments are switched on. In either case, a negligible DC voltage is applied across the segments.

The digital part of the ICL7107 is identical to the ICL7106 with the exception that the regulated supply and BP signal are not present and that the segment driver capacity has been increased from 2 to 8 mA. This current is typical for most LED seven-segment dis-plays. Since the driver of the most significant digit must take the current of two segments (pin 19), it has twice the current capacity of 16 mA.

Three methods can basically be used for a circuit of the clock generator:

- Using an external oscillator on pin 40
- Quartz between pin 39 and pin 40
- RC oscillator using pins 38, 39 and 40

The oscillator frequency is divided by four before it is used as the clock for the decade counters.

The oscillator frequency is then divided further down to derive the three-cycle phases. These are signal integration (1000 cycles), reference integration (0–2000 cycles), and automatic zeroing (1000–3000 cycles). For signals that are smaller than the input range end value, the unused part of the reference integration phase is used for automatic zeroing. This results in the total duration of a measurement cycle of 4000 (internal) clock periods (corresponds to 16,000 external clock periods), independent of the input voltage magni-tude. For about three measurements per second, a clock frequency of about 50 kHz is therefore used.

To obtain maximum suppression of the mains frequency components, the integration interval should be selected so that it corresponds to a multiple of the mains frequency period of 20 ms (at 50 Hz mains frequency). To achieve this property, clock frequencies of 200 kHz ($t_i = 20$ ms), 100 kHz ($t_i = 40$ ms), 50 kHz ($t_i = 80$ ms), or 40 kHz ($t_i = 100$ ms) should be selected. It should be noted that if a clock frequency of 40 kHz is selected, not only the mains frequency of 50 Hz is suppressed, but also 60, 400, and 440 Hz.

3.2.2   External Components of ICL7106 and ICL7107

The following external components are required to operate the ICL7106 and ICL7107:

3.2.2.1 Integration Resistance R_I

Both the input amplifier and the integration amplifier have a class A output stage with a quiescent current of 100 µA. They are capable of delivering current from 20 µA with negligible non-linearity. The integration resistance should be chosen high enough to remain in this very linear range for the entire input voltage range. On the other hand, it should be small enough so that the influence of unavoidable leakage currents on the circuit board does not become significant. For an input voltage range of 2 V value of 470 kΩ is recommended and for 200 mV one with 47 kΩ.

3.2.2.2 Integration Capacitor

The integration capacitor should be dimensioned in such a way that, taking into account its tolerances, the output of the integrator does not enter the saturation range. A value of 0.3 V should be maintained as the distance between the two operating voltages. When using the "internal reference" (ANALOG COMMON), a voltage swing of ±2 V at the integrator output is optimal. For ICL7107 with ±5 V operating voltage and "ANALOG COMMON" with reference to the operating voltage, this means that an amplitude of ±3.5 to ±4 V is possible. For three measurements per second the capacitance values 220 nF (7106) and 100 nF (7107) are recommended.

It is important that if other clock frequencies are selected, these values must be changed to achieve the same output voltage swing.

An additional requirement on the integration capacitor is low dielectric losses to minimize the "roll-over" error. Polypropylene capacitors give the best results here at a relatively low cost.

3.2.2.3 "AUTO-ZERO" Capacitor C_z

The value of the "AUTO-ZERO" capacitor influences the noise of the system. For an input voltage range end value of 200 mV, where low noise is very important, a value of 0.47 µF is recommended. In applications with an input voltage range end value of 2 V, this value can be reduced to 47 nF to reduce the recovery time of input overvoltage conditions.

3.2.2.4 Reference Capacitor C_{ref}

A value of 0.1 µF shows the best results in most applications. In those cases where a relatively high common-mode voltage is present, e.g. when "REF LOW" and "ANALOG COMMON" are not connected, a higher value must be selected for an input voltage range end value of 200 mV in order to avoid "roll-over" errors. A value of 1 µF would have a roll-over error of less than 1/2 digit in these cases.

3.2.2.5 Components of the Oscillator

For all frequencies, a resistor from 100 kΩ should be selected. The capacitor can be determined according to the function:

$$f = \frac{0.45}{R \cdot C}$$

A value of 100 pF results in a frequency of about 48 kHz.

3.2.2.6 Reference Voltage

To reach the range end value of 2000 internal clocks, an input voltage of $U_{IN} = 2\,U_{REF}$ must be applied. Therefore the reference voltage for 200 mV input voltage range must be selected to 100 mV, for 2000 mV input voltage range to 1000 mV.

3.2.2.7 Operating Voltages of the ICL7107

The ICL7107 is designed to work with operating voltages of ±5 V. However, if a negative supply is not available, it can be generated with two diodes, two capacitors, and a simple CMOS gate. In certain applications no negative operating voltage is necessary under the following conditions:

Condition one:	The reference of the input signal is in the middle of the common-mode voltage range
Condition two:	The signal is less than ±1.5 V

Voltage losses at the capacitors generate leakage currents. The typical leakage current of the internal analog switches (I_{DOFF}) at nominal operating voltages is 1 pA and 2 pA at the input of the input amplifier and the integrated amplifier. With regard to the offset voltage, the influence of the voltage drop on the "AUTO-ZERO" (capacitor and that of the drop on the reference capacitor) is opposite, i.e. no offset occurs if the voltage drop on both capacitors is the same. A typical value for the offset caused by this voltage drop in relation to the input results from a leakage current from 2 pA, which discharges a capacitance from 1 µF for 83 ms (10,000 clock periods at a clock frequency of 120 kHz) to an average value of 0.083 µV.

The effect of this voltage drop on the "roll-over" error (different numerical displays for the same positive and negative input values for input voltages close to the respective range end value) is slightly different. With negative input voltages, an analog switch is closed during the deintegration phase. Thus, the influence of the voltage drop on the reference capacitor and on the "AUTO-ZERO" capacitor is "differential" for the entire measuring cycle (and ideally compensates itself). For positive input voltages, an analog switch is closed in the deintegration phase and the "differential" compensation is no longer present in this phase. Here we get a typical value from 3 pA, which discharges 1 µF for 166 ms, to 0.249 µV.

These figures indicate that the source of error discussed in this section is irrelevant at 25 °C. At an ambient temperature of 100 °C, the corresponding values are 15 and 45 µV respectively. With a reference voltage of 1 V and a system counting up to 20,000, 45 µV corresponds to less than 0.5 of the least significant digit (but with a reference from 200 mV, the values are already four to five counters!)

Voltage changes on the capacitors do not cause charge over coupling with the switch-off edge of the switching control signals. It is no problem to charge the capacitors to the correct value with the analog switches turned on However when the switch is turned off, the GATE-DRAIN capacitance of the switch causes charge over coupling to the reference and "AUTO-ZERO" capacitors, changing the voltage applied to them. The charge over coupling caused by switching off the analog switch can be measured indirectly as follows: Instead of 1 µF, 10 nF is used as an "AUTO-ZERO" capacitor. In this case, the offset is typically 250 µV. If one now looks at the integration output voltage over time, the result is essentially a linear curve, which suggests that the relevant influence must be charge over coupling. If it were the leakage current, a quadratic dependence would result!

From the 250 µV results with $C = Q \cdot U$ an effective over a coupled charge of 2.5 pC or capacitance of 0.16 pF, with an amplitude of the gate control voltage of 15 V.

The influence of the internal five analog switches is more complicated, because—depending on the time—some switches are switched off while others are switched on. Using a reference capacitor from 10 nF instead of the nominal value from 1 µF results in an offset of less than 100 µV. Thus, the error caused by these charges over couplings is about 2.5 µV for a capacitor from 1 µF. It has no influence on the "roll-over" error and does not change significantly with temperature.

The external components are dimensioned for a measuring range of 200.0 mV and three measurements per second. "IN LOW" can either be connected to "COMMON" for "floating" inputs relative to the supply or to "GND" or "0 V" when the differential input is not used.

Since the signal voltage and the reference voltage are fed into the same input of the circuit with the input amplifier, the amplification of the input amplifier and the integrator amplifier has no significant influence on the accuracy in the first approximation, i.e. the input amplifier can have a very unfavorable common-mode rejection over the input voltage range and still not cause an error as long as the offset voltage changes linearly with the input common-mode voltage.

The first cause of error here is the non-linear term of the common-mode voltage suppression.

Careful measurements of the common-mode voltage suppression on 30 amplifiers showed that the "roll-over" error is possible from 5 to 30 µV. In any case, the error due to the non-linearity of the integrator is less than 1 µV.

When the input is short-circuited, the output of the input amplifier goes to U_{ref} (1 V) in 0.5 µs with an approximately linear progression. Thus 0.25 µs of the deintegration time is lost. For a clock from 120 kHz, this means about 3% of the clock period or 3 µV. There is no offset error because this delay is the same for positive and negative reference voltages. The converter switches from 0- to 1-signal at the input at 97 instead of at 100 µV.

The comparator with 3 µs introduces a much greater delay in the circuit. At first glance, this seems to be a small value, comparing 3 µs with the 10–30 ns of some comparators. However, the latter are specified for overdriving from 2 to 10 mV. If the comparator at the input has an overdrive from 10 mV, the zero-crossing of the integrator output is already some clock periods behind!

The comparator used has a gain-bandwidth product from 30 MHz and is therefore comparable to the best-integrated comparators. The only problem is that it has to work with 30 µV instead of overdrive from 10 mV. The switching delay of the comparator does not cause an offset but causes the converter to switch from 0 to 1 signal at 60 µV, from 1 to 2 at 160 µV, etc. For most users, this switching at approx. 1/2 LSB is more pleasant than the so-called "ideal case" in which switching is performed at 100 µV.

If it is nevertheless necessary to get close to the "ideal case", the delay of the comparator can be approximately compensated by connecting a small resistor value (approx. 20 Ω) in series with the integration capacitor. The time delay of the integrator is at 200 ns and does not contribute to any measurable error.

3.2.3   Integrating AD Converters with the ICL7106 and ICL7107

Every integrating A/D converter assumes that the voltage change at a capacitance is proportional to the time integral of the capacitor current.

$$C \cdot \Delta U_C = \int i_C(t)\, dt.$$

In reality, however, a very small percentage of charge is "misused" for charge rearrangement within the dielectricum of the capacitor. These charge components naturally do not contribute to the voltage at the capacitor and this effect is called dielectric loss.

Probably one of the most accurate methods of measuring dielectric losses of a capacitor is to use it as an integration capacitor in an integrating A/D converter, with the reference voltage applied as input voltage (ratiometric measurement). The ideal value on the display would be 1.0000, independent of the values of the other components. Very careful measurements under the observation of the zero crossings in order to be able to extrapolate to a fifth digit and mathematical consideration of all delay errors resulted in the following display values for different dielectrics:

Dielectric	Display
Polypropylene	0.99998
Polycarbonate	0.9992
Polystyrene	0.9997

As a result, polypropylene capacitors are very well suited for this application. They are not very expensive and the relatively high-temperature coefficient has no influence. The dielectric losses of the "auto-zero" and reference capacitors only play a role when the operating voltage is switched on or when "returning" from an overload condition.

Normally the external reference of 1.2 V is connected to "IN LOW" with "COMMON" to set the correct common-mode voltage. If "COMMON" is not connected to "GND", the input voltage can "float" relative to the operating voltages, and "COMMON" acts as a pre-regulation for the reference. If "COMMON" is short-circuited with "GND", the differential input is not used and the pre-control is ineffective.

Apart from leakage currents and overcoupling switching edges, charge losses at the reference capacitor can also be caused by capacitive voltage division with a stray capacitance c_S (capacitance before the buffer). An error only occurs with positive input voltages.

During the "auto-zero" phase both capacitors, C_{ref} and C_S are charged to the reference voltage via the analog switch. If a negative input signal is now applied, C_{ref} and C_S are in series and form—with respect to C_{ref}—a capacitive voltage divider. For $C_S = 15$ pF the divider ratio is 0.999985.

If the positive reference is now switched to the input via the analog switch in the deintegration phase, the same voltage divider is in action as in the signal integration phase. If both voltage integration and reference integration work with the same divider, no error is caused by this divider.

For positive input voltages, the divider is active in the signal integration phase in the same way as for negative input voltages. The negative reference is switched on at the beginning of the deintegration phase by closing the analog switch. The reference capacitor is not used and the divider is not in action. In this case, the corresponding divider ratio is 1.0000 instead of 0.999985.

This error, which depends on the input voltage, has a gradient of 15 µV/V and results in a "roll-over" error of 30 µV, i.e. h., the negative display is too low by 30 µV.

When implementing an integrating AD converter ICL7106 and ICL7107, four types of errors must be considered. With the recommended components and a reference voltage from 1 V, these are

- Offset error of 2.5 µV due to charge overcoupling of switching edges
- A "roll-over" error from 30 µV at the end of range caused by the stray capacitance c_S
- A "roll-over" error from 5 to 30 µV at full scale due to non-linearity of the input amplifier
- A "delay error" from 40 µV when switching from 0 to 1 signal

The values correspond well with the actual measurements. Since the noise is about 20 µVss, only the statement that all offset voltages are smaller than 10 µV is possible. The observed "roll-over" error corresponds to half a counter (50 µV), where the negative indication is larger than the positive one. Finally, switching from 0000 to 0001 is done with an input voltage of 50 µV. These figures show the performance of a reasonably designed integrating AD converter, although it should be noted that these data are achieved without particularly accurate and thus expensive components.

Due to a delay of 3 µs of the comparator, the maximum recommended clock frequency of the circuit 160 kHz. In the error analysis it has been shown that in this case half of the first clock period of the reference integration cycle is lost, i.e. the display goes from 0 to 1 at 50 µV, from 1 to 2 at 150 µV, etc. As mentioned before, this feature is desirable for many applications.

However, if the clock frequency is increased significantly, the display will change in the last digit even if the input is short-circuited due to noise peaks.

The clock frequency can be selected higher than 160 kHz by connecting a small resistance value in series with the integration capacitor. This resistor causes a small voltage jump at the output of the integrator at the beginning of the reference integration phase.

By carefully choosing the ratio of this resistor to the integration resistor (recommended are 20–30 Ω), the delay of the comparator can be compensated and the maximum clock frequency can be increased to approx. 500 kHz (corresponding to a conversion time of 80 ms). At even higher clock frequencies, the circuit is considerably restricted by frequency response restrictions in the range of low input voltages.

The noise figure is approximately 20 µVss (*3σ* value). Near the end of the measuring range, it increases to approx. 40 µV. Since much of the noise is generated in the "auto-zero" feedback loop, the noise behavior can be improved by increasing the input amplifier gain by about five. Increasing the gain will cause the auto-zero switch to no longer switch through properly due to the increased offset voltage of the input amplifier.

In many applications, the secret of a system's performance lies in the correct use of the individual components. The A/D converter can also be considered as a single component of a system, and therefore a reasonable design of the system is necessary to achieve optimum accuracy. The monolithic AD converters are very accurate due to the integration method used. In order to use them optimally, the circuit design and the selection of the external passive components should be done with the necessary care. The measuring instruments used should be much more accurate and stable than the system to be developed.

The wiring of the reference potential must be planned thoroughly because it is important to avoid "ground loops". According to all experience, the most frequent cause of faults in an AD system is unfavorable wiring of the reference potential. The operating currents of the analog section, the digital section, and the display all flow through one connection—the reference for the analog input.

The average value of the current flowing through the reference terminal of the input generates an offset voltage. Even the automatic zeroing circuit of an integrating converter is not able to compensate for this offset. In addition, this current has some alternating components. The clock generator and the various digital circuits that are driven produce alternating current components with the clock frequency and possibly "subharmonics" of this frequency. In the case of a converter with successive approximation, this produces an additional offset. With an integrating converter, at least the higher frequency components should be averaged out.

With some converters, the analog operating currents also change with the clock or a "subharmonic" of it. If the display is operated in multiplex, this current changes with the multiplex frequency, which is normally derived by lowering the clock frequency. In an integrating converter, the currents of the analog part and the digital part will differ for the different conversion phases.

Another important reason for the change in operating current is that the operating currents of the digital part and the display depend on the measured value displayed. This is often expressed by the flickering of the display and/or by missing measured values. A displayed value changes the effective input voltage (by changing its reference potential).

This causes a new measured value to be displayed, which again changes the effective input voltage, etc. This then causes the display to oscillate between two or three values despite a constant voltage at the input of the system.

Another potential source of error is the clock generator. If the clock frequency changes due to changes in operating voltage or current during a conversion cycle, inaccurate results are obtained.

The digital and analog reference lines are connected by a line through which only the compensating current flows between these parts. The display current does not influence the analog part and the clock part is blocked by a decoupling capacitor. It should be noted that the currents of any external reference used, as well as any further current from the analog section, must be carefully fed back to the analog reference.

After setting up the circuit and visually checking for errors, switch on the operating voltage. Depending on the grinding position of the spindle adjuster, some value will appear on the display. If the seven-segment displays do not light up or if the adjustment described below cannot be carried out, the operating voltage must be switched off immediately and the circuit checked again.

To calibrate the zero point, hold the sensor in ice water and set the display to "00.0" with the spindle trimmer P_2. To do this, half fill a glass of water with crushed ice cubes, add a little water until about half the height of the ice pieces is covered. Afterward, use the spindle trimmer to set the display to exactly "00.0".

3.3 Thermocouples

Thermocouples are inexpensive and robust, and unlike most other temperature sensors, they have relatively good long-term stability. Furthermore, their small external dimensions and fast response behavior over wide temperature ranges are often decisive for their use. Applications range from low-temperature technology to measurements on jet engines or in blast furnaces. The accuracy is quite good, the characteristic curve is not exactly ideal, but can be linearized.

3.3.1 Thermoelectric Effect

The number of free electrons in a metal depends on the temperature and on its composition. The thermoelectric effect was discovered by Seebeck in 1822, hence the term "Seebeck" effect. Already in 1826 A.E. Becquerel used a platinum-palladium element to measure temperature. If two different metals are joined together, a potential difference arises. The voltage generated is a function of temperature and is generally very small. Table 3.2 lists some of the most common thermocouples.

Each thermocouple consists of two different metals. A thermocouple of copper and solder generates a voltage of 3 µV/K. Since an electrical circuit always consists of at least

Table 3.2 Temperature ranges and thermoelectric voltage for the most important thermocouples according to DIN IEC 584

Thermocouple, material	Type	Measuring range (°C)	Thermoelectric voltage (mV)
Cu-CuNi	T	−270 − +400	−6.26–20.87
NiCr-CuNi	E	−270 − +1000	−9.84–76.36
Fe-CuNi (Fe-Const)	J	−210 − +1200	−8.1–69.54
NiCr-Ni	K	−270 − +1372	−6.64–54.88
PtRh 13-Pt	R	−50 − +1769	−0.23–21.10
PtRh 10-Pt	S	−50 − +1769	−0.24–18.69
PtRh 30-Pt-PtRh 6	B	0 − +1820	0.00–13.81

two contacts in series, care must be taken when measuring with thermocouples to ensure that such unwanted and random thermocouples do not cause corresponding measuring errors. The selection criteria for thermocouples are

3.3.1.1 Fe-CuNi
has an almost unlimited service life at temperatures up to 500 °C, but above 600 °C, the Fe wire begins to scale strongly. This thermocouple is very resistant to reducing gases, except hydrogen. The disadvantages are so great that it is hardly used in practice.

3.3.1.2 Cu-CuNi
has the advantage over Fe-CuNi that it does not rust. Since copper oxidizes from 400 °C, the range of applications for this thermocouple is somewhat limited. This thermocouple is used very often in practice because it is very inexpensive but at risk of corrosion.

3.3.1.3 NiCr-Ni
is most resistant to oxidizing gases, but is particularly sensitive to sulfur-containing gases and is attacked by silicon vapors in reducing atmospheres. Gas mixtures with an oxygen content under 1% cause "green rot", i.e. the thermoelectric voltage and the strength change. In practice, this thermocouple is very often used in the measuring range of 800 to 1000 °C. This group also belongs the NiCr-CuNi thermocouple, which generates a high thermoelectric emf, but is nevertheless hardly used. The NiCrSi-NiSi thermocouple can also be classified in this category, but in practice, it is not very common. The working range extends to 1300 °C; this type can partially replace more noble and thus expensive thermocouples.

3.3.1.4 PtRh10-Pt
is particularly susceptible to contamination of any kind due to the purity of the metals used, but this thermocouple offers good chemical resistance in oxidizing atmospheres. The costs are very high, but it offers very good long-term stability and a low tolerance for production.

Fig. 3.23 Block diagram of a measuring circuit with a thermocouple, with an ice bath of 0 °C to generate the reference temperature

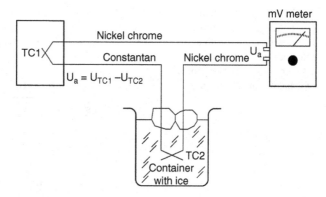

3.3.1.5 PtRh30-PtRh6

has similar corrosion-chemical properties, but is somewhat less sensitive to contamination. The cost is very high, but it is suitable for extreme temperatures. However, this thermocouple has the lowest thermoelectric voltage.

Figure 3.23 shows a typical thermocouple application with measuring junction TC1 and reference junction TC2. The thermocouple at TC1 consists of a nickel-chromium wire and a constant and wire, resulting in the NiCr-CuNi thermocouple. The same applies to the thermocouple at TC2.

3.3.2   Measurements with Thermocouples

In order to be able to use the thermoelectric emf as a measure of temperature, the free ends of the thermocouple must be at a constant reference temperature in the reference junction—for example, in a vessel with ice water of 0 °C. Since it might be a bit of a hindrance to always carry an ice bucket around with you, you can also do without the ice bath and instead measure the temperature of the reference junction, as shown in Fig. 3.24.

If the reference junction is at a temperature of +25 °C, for example, the reference junction thermocouple subtracts a thermoelectric emf corresponding to 25 °C. Therefore this voltage must be added again at another point. This method seems a bit nonsensical at first because now a second measurement has to be made. Since you can determine the temperature of the reference junction yourself and it is normally located at or in the measuring instrument, you can assume that the temperature is in the range from −20 to +70 °C. In this range, it can be measured without any problems with a semiconductor sensor.

Although thermocouples have a low output impedance, they provide a very low voltage. Therefore, subsequent signal processing is not very simple. Useful signals of only a few millivolts require complex follow-up electronics with very low drift if temperature resolutions in the order of 1 K are required. For most types, the linearity is not very good. However, since the characteristics are precisely known, linearization can be carried out either analog or digital in the course of further signal processing.

Fig. 3.24 Practical example
of temperature measurement
for a thermocouple with a
reference junction, where the
different surface temperatures
of a candle can be measured

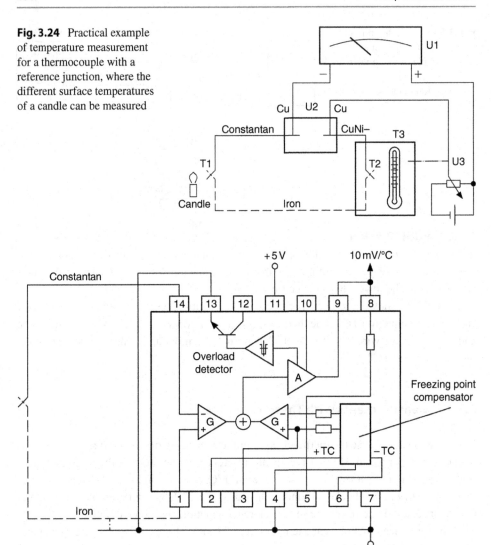

Fig. 3.25 Measuring circuit for thermocouples with the AD594. The temperature range is between
0 and +300 °C

Until 1985, the operation of thermocouples required quite sophisticated amplifier electronics. Since then, the AD594 device (Fig. 3.25) has brought about a considerable simpli-
fication. It contains instrumentation amplifiers and compensation circuitry on one chip.
This means that the second thermocouple for compensation can be omitted. The AD594
uses the two parasitic elements of the terminals for this purpose. Since three different metals (iron/copper, constantane/copper or nickel/copper, chromium-nickel/copper) are
involved in this case, the thermoelectric voltages do not cancel each other out even at low

temperatures at the terminals. The result is a thermocouple whose output voltage is too small by the difference between the two terminal voltages. This differential voltage must then be added to the actual measuring voltage, as in the previously discussed method.

3.3.3 Amplifier for Thermocouples

In conjunction with an ice point reference, an already calibrated amplifier reaches an output sensitivity of 10 mV/K. Depending on the external circuitry, this amplifier thus serves as a compensator or as a switch with externally adjustable threshold values. It is possible to obtain the compensation voltage of this amplifier directly as an output signal. In this variant, the integrated circuit itself is the element which detects the temperature. In addition, there is an alarm circuit that triggers a signal if there is an interruption in the thermocouple cable. This is obtained by switching on an LED with a series resistor between the operating voltage and pin 12. This results in a visual indication of the alarm status. Instead of supplying the LED, this output can also provide a control signal with a TTL level if a pull-up resistor is connected; this can be used to trigger an interrupt in the microprocessor or microcontroller. This allows quick detection of the measurement error.

The AD594 can be operated with only a single voltage from +5 V, with the temperature range to be measured limited to values between 0 and +300 °C. However, it also operates at a double operating voltage of up to ±15 V by disconnecting the connection between pin 4 and pin 7. Pin 4 now forms the ground connection, the negative operating voltage is connected to pin 7, the positive operating voltage remains at pin 11. Asymmetrical voltage is important for this operating mode. This enables a temperature range of −184 and +1260 °C to be achieved, whereby the entire range of a thermocouple can be fully utilized.

Each chip is adjusted by laser; the AD594 according to the characteristic curve of an iron-constantan thermocouple. Figure 3.26 shows the deviations of the measurement results. The AD595 is matched to nickel-chromium/nickel thermocouples. With J and K thermocouples (AD594 or AD595), no external components or adjustment work is required. Other versions of thermocouples can be easily matched using external resistors.

However, the AD594/595 cannot compensate for one characteristic of thermocouples: linearity error. As long as very high precision is not required, this does not play too great a role in the range of a few hundred °C. In contrast, linearization is necessary for high accuracy or large measuring ranges. In many cases, this problem is solved by software on the PC. If necessary, a precision analog multiplier can also be used for this purpose.

Linearization can be dispensed with if the reproducibility of a temperature range is the only important factor, such as in the automatic control of a thermal process, but not cold junction compensation. In the vast majority of cases, the change in ambient temperature at the reference junction is so large that it causes unacceptably high errors at the output of the thermocouple pair. To prevent this, either the reference junction must be kept at a constant temperature (ice bath or thermostatically controlled oven) or the voltage generated at the cold junction must be compensated due to the changes in ambient temperature. Since cold

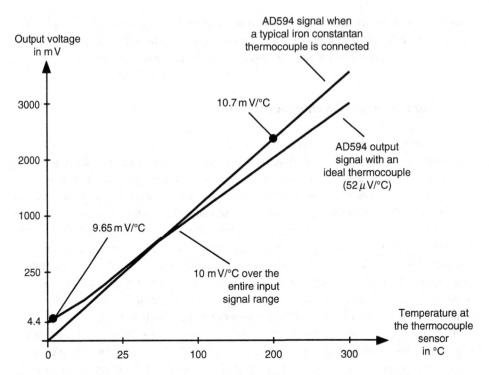

Fig. 3.26 Deviations in the measurement results of an iron constantan thermocouple in conjunction with the AD594

junction compensation is integrated into the AD594/595, no additional circuitry effort is required. Low power consumption and low thermal resistance of the housing guarantee that the error due to self-heating remains negligible. In still air, the thermal resistance from the IC to the environment is only 80 K/W. The low power consumption of only 800 µW results in self-heating in the air of less than 0.065 K.

When setting up the mechanical structure, special care must be taken that the thermo-couple is connected as directly as possible to the AD594/595, since the temperature of these contact points (reference junction) must be the same as that registered by the tem-perature sensor on the chip. The use of a socket is therefore not recommended. The inter-position of extension cables, plugs, switches, or relays is only permitted if the conductor materials are identical to the metals of the thermocouple. Otherwise parasitic thermoelec-tric voltages will inevitably be generated which are not compensated.

3.4   Resistance Thermometers with Pt100 or Ni100

In terms of its sensitivity (Ω/K), nickel is clearly superior to platinum. Since the constancy and reproducibility of the characteristic curve can also be described as good, standardized Ni100-DIN resistance thermometers are often found in practice alongside the likewise standardized Pt100 types. Table 3.3 lists the most important properties.

Table 3.3 Comparison between Pt100 and Ni100

	Pt100	Ni100
Measuring range °C	−200 − +850	−60 − +180
Resistance change Ω/K	0.42–0.32	0.47–0.81

Fig. 3.27 Sensitivity (slope) of Pt100 and Ni100 resistance thermometers

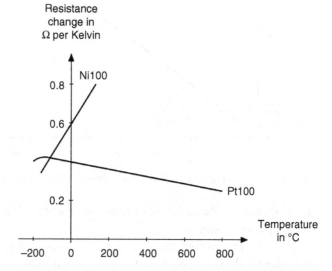

Figure 3.27 compares the steepness of the characteristic curves of Pt100 and Ni100 resistance thermometers. The characteristic curve of platinum decreases as the temperature rises, i.e. the sensitivity decreases, whereas that of nickel increases, the sensitivity increases. A sensor with linear characteristics would have a horizontal straight line in this illustration. The number 100 in the standard designation indicates that R_0 at 0 °C is chosen as 100 Ω.

3.4.1 Pt100 Resistance Thermometer

In contrast to a thermocouple, a Pt100 resistance thermometer is a passive element, i.e. the mode of operation is based only on the temperature dependence of the electrical resistance. The temperature measurement is therefore a pure resistance measurement, whereby an optimum condition is obtained in conjunction with the standardized value of the Pt100 and its reference temperature.

Depending on the design, the platinum measuring resistor is embedded as a platinum wire or strip in a ceramic or glass body or is located as a thin layer on a ceramic plate Fig. 3.28 shows the characteristic curve. The connecting wires of the measuring element are connected to the active resistor part in a vibration-proof way. The matching connections of multiple measuring resistors differ in the length of the connecting wires. As an

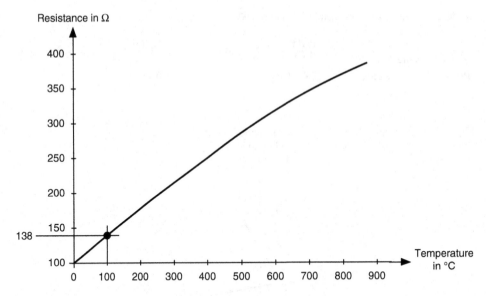

Fig. 3.28   Characteristic curve of the Pt100 resistance thermometer

extended element, the measuring resistor records the average value of the temperature prevailing over its length.

All Pt100 versions correspond in their basic values and limit deviations to the standard DIN IEC751. The DIN specifications for nominal resistance values from 100 Ω apply. For resistors with n times, the nominal value of 100 Ω the basic values or limiting deviations must also be multiplied by n.

The permissible measuring current depends mainly on the thermal contact between the measuring resistor and the medium whose temperature is to be measured. The type of medium is also relevant. For example, with a resistance thermometer used in flowing water, a considerably larger measuring current is possible with the same heating error than with the same sensor in air.

Since in practice measurements are carried out under completely different conditions, theoretical recommendations for the measuring current must be omitted here. In the datasheets, the self-heating coefficient "S" in Kelvin per milliwatt of power consumed is given for each individual measuring resistor. For a given measuring current, the power can be calculated using the basic value series with

$$P = I^2 \times R$$

can be calculated. According to the equation

$$\Delta T = P \times S$$

then the self-heating error ΔT in Kelvin results.

The Pt100 measuring resistors can be used for direct and alternating current measurements. The glass versions G and GX as well as the layer measuring resistors are practically induction-free, for the ceramic resistors K, KE, and KN a low inductance is possible, but with max. 100 µH is meaningless.

The half-life is the time it takes a thermometer to detect half of a temperature jump. The 9/10 time is defined analogously. These two response times are specified for water at 0.4 m/s flow velocity and for air at 1 m/s and can be converted to any medium with a known heat transfer coefficient according to VDI/VDE 3522. With regard to resistance thermometers, one speaks of hysteresis if the resistance values at certain temperatures are different from the initial state after a temperature cycle (e.g. cooling and reheating). Likewise, hysteresis is characterized by the fact that the changes in measured values can be made to disappear again or overcompensated by a contrary temperature cycle. This process is therefore reversible. In the case of resistance thermometers, the measured value hysteresis can occur after shock-like temperature changes.

For a reliable temperature measurement with platinum measuring resistors, it is necessary that the mechanical characteristics of the measuring resistor (size, shape, vibration resistance, temperature range, response time, insulation resistance, and other functions) are adapted to the measuring task and the conditions at the measuring location.

Besides the selection of the measuring resistor type, the installation at the measuring location is therefore of great importance. It is not possible to give generally valid advice, a high degree of experience is necessary. In practice, it is recommended to consult a specialist.

In practice, not only the Pt100 but also the Pt500 and Pt1000 can be found. At 0 °C, these have a basic resistance of 500 or 1000 Ω.

3.4.2   Ni100 Resistance Thermometer

Instead of platinum, nickel can also be used for a resistance thermometer. It is cheaper and has a temperature coefficient twice as high. However, the measuring range is somewhat limited and ranges from −60 to +250 °C. Figure 3.29 shows the characteristic curve for a Ni100.

The Ni100 resistance thermometer has a resistance value of 100 Ω at a measuring temperature of 0 °C with a permissible deviation of 0.2 Ω.

3.4.3   Silicon Temperature Sensor as Pt100 Replacement

Until 1990, only semiconductor temperature sensors were offered on the sensor market, which was defined at a temperature of mostly 25 °C with a tolerance specification. Due to a new manufacturing process and a complex computer-aided measuring system, it is now possible to specify the KTY87 series of Si-temperature sensors with a low scattering of

Resistance in Ω

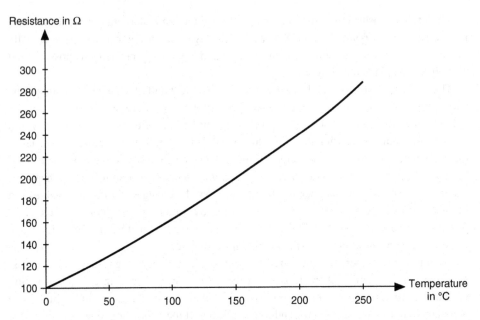

Fig. 3.29 Characteristic curve of a Ni100 resistance thermometer

±0.5% at temperatures from 25 to 100 °C. This ensures that the sensor does not exceed a temperature error of ±0.8 K between 20 and 100 °C. If one compares the KTY87 family with the Pt100 series, there is no significant difference in the properties, but mainly in the price.

Due to component tolerances and the offset voltage of the standard operational amplifiers, the evaluation circuit for the temperature sensor must be adjusted to take full advantage of its measuring accuracy. However, the absolute measuring accuracy of the KTY87 in the range from 25 to 100 °C allows the amplifier to be calibrated in relation to the nominal resistance values of the KTY87 and then operated with any sensor copy, while keeping the temperature measuring error smaller than ±1 K. In the temperature range from 20 to 100 °C, the KTY87 is an inexpensive alternative to the Pt100 temperature sensor if the slightly lower measuring accuracy is accepted.

In the circuit shown in Fig. 3.30, the operational amplifier operates as a differential amplifier with an operating voltage of +5 V. This circuit can therefore be used for direct control of an AD converter. Here, metal film resistors with tolerance below ±0.5% and a 1-K value below ±50 ppm/K are absolutely necessary. To calibrate the circuit, the sensor is replaced by a measuring resistor from 1640 Ω (nominal value of the KTY87 at 0 °C), and the output voltage is set to $U_a = 0.5$ V with P_1. Then a measuring resistor from 3344 Ω (nominal value at 100 °C) is used instead of the sensor and P_2 is used to set the output voltage to $U_a = 4.5$ V. The circuit for the nominal temperature sensor KTY 87 is thus adjusted. This circuit can be used with any KTY87 sensor, whereby the remaining measurement error of less than ±1 K is maintained in the range 20–100 °C,

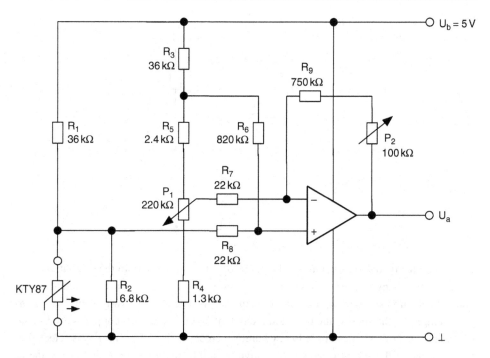

Fig. 3.30 Evaluation amplifier for the KTY87 temperature sensor

If the amplifier is to be adapted to a specific sensor, the adjustment procedure is carried out with this sensor instead of with measuring resistors, which is then exposed to the temperatures of the adjustment points, e.g. in a precise liquid thermostat. This considerably more complex method brings a further gain in measuring accuracy, but only for this specific sector on the entire characteristic curve.

3.4.4   Connection of a Resistance Thermometer

Resistance thermometers can be connected according to three methods:

- Two-wire circuit,
- Three-wire circuit,
- Four-wire circuit.

Figure 3.31 shows the classical method for the two-wire circuit. In front of the operational amplifier, there is a resistance bridge, the left branch of which contains the sensor and adjusters, while the right branch consists of fixed resistance values.

With the temperature sensor, one must essentially only pay attention to the self-heating so that no measurement distortion occurs. The circuitry required for this variant is very

Fig. 3.31 Two-wire circuit for
a resistance thermometer

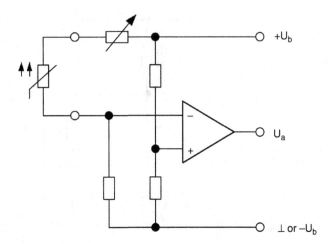

low, but there are also disadvantages if the sensor is too far away from the evaluation electronics. Any change in temperature on the supply line to the sensor is included in the complete set up as a measurement error. If the sensor is e.g. 25 m away from the operational amplifier, the supply line has a total length of 50 m. In practice, copper is used as the conductor material for the supply line, whereby a cross-section of 0.5 mm² was selected in this example. If the sensor operates in a range from 0 to +100 °C, the following temperature dependence on the supply line results:

$$R_{\mathrm{L}} = R_{\mathrm{K}} = \frac{l \cdot \rho}{A} = \frac{25\mathrm{m} \cdot 0.01724\Omega \cdot \mathrm{mm}^2 \,/\, \mathrm{m}}{0.5\mathrm{mm}^2} = 1.724\Omega$$

$$R_{\mathrm{W}} = R_{\mathrm{K}} \left(1 + \alpha \times \Delta T\right) = 1.724\Omega \; \left(1 + 0.00393 \,/\, \mathrm{K} \times 100\mathrm{K}\right) = 2.4\Omega$$

The resistance change of 2.4 Ω corresponds to an error of about 6 K if the circuit was optimally adjusted at 20 °C.

To keep the influence of the line resistances on the evaluation electronics as low as possible, the three-wire circuit of Fig. 3.32 is used. The temperature sensor receives a direct power supply, which simultaneously operates the left branch of the bridge circuit. This eliminates temperature fluctuations on the measuring lines since the left branch is also fed back to the evaluation electronics. Any kind of temperature change on the measuring lines can be compensated by the three-wire circuit.

The four-wire circuit shown in Fig. 3.33 is ideal for measurement technology. The measuring resistor is located between the two inputs of the operational amplifier and can therefore act directly. The problem, however, is the operating state of the measuring resistor, which must be driven with a constant current, i.e. a constant current source must be connected in the supply line to the measuring resistor.

Fig. 3.32 Three-wire circuit
for a resistance thermometer

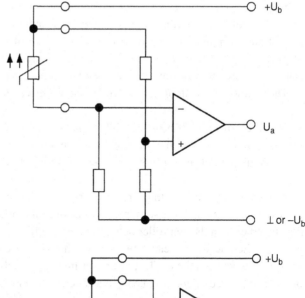

Fig. 3.33 Four-wire circuit
for a resistance thermometer

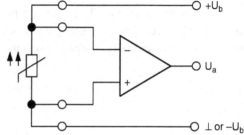

3.4.5   Avoidance of Electromagnetic Susceptibility

Electrostatic and capacitive radiation of interference into a device is relatively unlikely
with aluminum housings. However, magnetic irradiation is possible. These can be generated by inductive components, especially in conjunction with thyristors or TRIACs.
Closed cores (transformers, chokes) generate much smaller stray fields than open ones
(relays, contactors). The stray field is more pronounced in the longitudinal direction of the
coil than transverse to it. While non-ferrous metals (including some stainless steels such
as "V4A") do not provide a shielding effect, the magnetic fields of steel plates are fully
absorbed in thicknesses as low as 2–20 mm. The most effective magnetic shielding is provided by "Mu-Metal".

Interference caused by magnetic fields only occurs when the field strength changes
over time. With coils through which direct current flows, the magnetic field changes only
at the moment of switching on and off. However, due to the usually higher number of turns
and the non-closed core, the stray field generated by this can exceed that of a coil through
which alternating current flows many times over. For the strength of the magnetic field,
only the strength of the flowing current is important, not the level of the applied voltage

(which, however, influences the current). Furthermore, the magnetic field increases with the number of turns and the length of the coil. Because of the inductance, coils for direct current have a higher number of turns than those for alternating current, with otherwise identical values. Therefore, small direct current relays with a very high number of turns can cause more intensive interference than large contactors with comparatively few turns.

The following installation instructions can be derived from this:

- As large a distance as possible between the test leads and relays, transformers, etc.
- Install inductors in such a way that they do not point in the longitudinal direction of the coil body towards measuring lines sensitive to interference.

Of greater importance than magnetic radiation from coils are those from adjacent cables, for example into the sensor lines of a controller, which are laid in a common cable duct together with the controller output lines as in Fig. 3.34. These couplings occur in two ways: on the one hand because two parallel lines represent a capacitor that can transmit interference between them. The power transmitted depends on the capacity and frequency of the interference; higher frequency interference is transmitted better than low-frequency interference. The intensity of transmission decreases with the distance between the lines

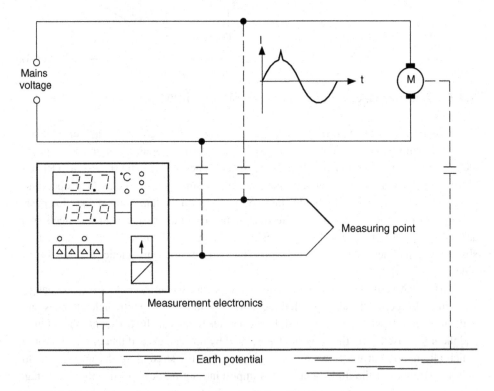

Fig. 3.34 Electromagnetic couplings on sensor and signal lines

and increases with the length. Theoretically, no interference is coupled into shielded cables, since their interior, as a "Faraday" cage, is free from the external interference field. However, the charges on the shielding must be able to flow off, which is why the shielding must always be connected to earth potential.

Secondly, interference is also transferred by magnetic means. The magnetic field enclosed by a pair of conductors forms between the forward and return lines and is the greater the distance between them. This applies both to interference fields generated by control lines and to the sensor lines. Lines through which alternating current flows permanently generate interference fields. A short-term, but considerable field is generated both bylines carrying direct and alternating current when switching inductive loads, e.g. of connected relays, as already described. It is also true that the induction voltage generated and thus the field change when a coil is switched off is several times greater than when it is switched on; the interference generated as a result is always much higher. When operating with DC voltage, it is, therefore, essential to use free-wheeling diodes, even if the coil is not switched by a transistor.

A practical solution is the external wiring of inputs with RC elements (resistor and capacitor), which then act as a low-pass filter and thus filter out interference, which is usually of a higher frequency nature. However, this causes problems, especially with controllers. Such a filter is a time element that increases the order of the controlled system by 1, thus making control more difficult. Such solutions are therefore only advisable in extreme cases.

Interference is not only coupled in via the input (signal) lines but also via the output (relay) lines. Therefore all lines, including, for example, those for external contacts, should be twisted. The distance between the different pairs of lines seems to have a more undulating effect than twisting.

The following installation instructions must therefore be observed:

- Lead sensor cables at a large distance, not parallel to supply and signal lines;
- Do not lay lines that are switched via the inductors parallel to the input and output lines;
- Use shielded and twisted cables for sensor cables, the shields must be earthed or connected to the housing;
- For signal and supply lines, run the outgoing and return lines as close as possible to each other and twist them if possible;
- For switched inductors (solenoid valves, relays, contactors) always install spark quenching combinations in the immediate vicinity of the inductor;
- Do not supply other devices, especially inductive loads, from the power terminals of a device, but use star-shaped wiring.

An unfavorable cable routing between the device and the sensor and the heating is shown in Fig. 3.35; here supply, output, and sensor cables are routed in parallel. In addition, the outgoing and return lines show large distances; lines for the supply of the load circuit have been connected to the power terminal of the device.

Fig. 3.35 Unfavourable cable routing from the device to the sensor and to the heater. Grounding and shielding are not drawn

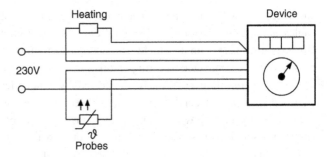

Fig. 3.36 Improved cable routing between device, sensor, and heater. Grounding and shielding are not drawn

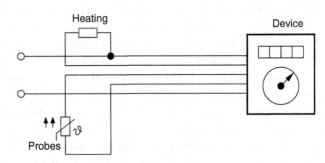

Figure 3.36 shows an improved cable routing with the sensor cables at a large distance from the current-carrying cables. Forward and return lines are laid in parallel at a small distance, and no other devices are supplied from the power terminals of the device.

The supply lines can also be disturbed—by short- or long-term overvoltages or short-term power failures. Longer-term overvoltages are most easily countered with varistors that conduct above a certain threshold voltage and short-circuit the overvoltage. In devices with primary switched-mode power supplies, such varistors are usually already integrated into the device. In devices with secondary switched-mode power supplies or simple power supplies with longitudinal transistors, overvoltages do not have such a drastic effect, as these are also transformed down—apart from lightning strikes or similar.

Short-term overvoltages, so-called transients, are caused by inductive devices such as motors, contractors, and the like connected to the same mains supply. They have short rise times and thus a high-frequency character. They can be filtered out with LC combinations. Such line filters are available in a wide range, they are inserted into the supply line near the device. Interference is diverted to the ground potential. It is therefore of great importance when installing such filters that they are properly grounded. On the other hand, the use of line filters can load the earth line with interference pulses, which can result in a "contaminated" earth potential.

The correct choice of filter requires a sound knowledge of the frequency band of the expected interference and, above all, the impedances of the network and the equipment. An incorrectly matched or insufficiently grounded filter makes the susceptibility to interference even worse, which is why they use of mains filters does not bring the desired success in many cases.

In order that no interference voltage can be induced on a shield, it must be discharged. In practice, a general reference potential, the "HF earth", is used for this purpose. This potential is always zero by definition. The PE (Protected Earth) line of the supply network is directly connected to the earth, for example by burying copper strips ("foundation grounding") and connecting them to the PE line on a so-called equipotential bonding rail. At transfer points where the consumer is connected to the power supply company (EVU), this earth line is connected to the neutral conductor N of the three-phase network arriving from the EVU. The neutral conductor is the return line for the operating currents of all loads connected between the external conductors L1, L2, L3, and N. Normally, no currents flow through the PE earthing conductor. For this reason, the protective conductor PE and the neutral conductor N should only be connected together at the transition point and in the immediate vicinity of the foundation earth electrode. If these are connected again at another location (the so-called "zeroing"), the grounding conductor can no longer meet the requirement for potential-free operation, since the neutral conductor does not necessarily have zero potential.

3.4.6 Ground Loops, Earthing and Shielded Cables

Nevertheless, potentials can also occur on an earth line, which prevents the charges from flowing off and cause various types of interference despite shielding. This is always the case when voltage sources and sinks occur on the ground line, which is often the case when several devices are connected to one ground line. It is therefore of elementary importance to provide each device with its own ground line to a common ground point.

Similar ground loops can also occur, for example, if switch cabinet parts—especially their doors—do not have their own grounding. Charges that occur on the cabinet door then flow out via the housing of a built-in controller and via its earthing cable. As a result, the controller housing is no longer at zero potential and the shielding is incomplete. It is also important to note that charges accumulate on protective conductors that are routed meter by meter parallel to supply lines, as is almost always the case with normal power lines, thus creating an unwanted potential.

Ground loops are caused by the fact that a current loop is created by unfavorable cable routing. This then acts like an air-core coil in which magnetic fields generated by interference sources can induce a current. Ground loops can be avoided by using separate ground lines since this considerably reduces the diameter of the conductor loop and thus the coupled field.

A question that comes up, again and again, is whether a shielded cable should be earthed at one or both ends. Certainly, it makes more sense to ground on both sides, as this allows the charges to be dissipated optimally. However, the problem usually arises that the earthing potentials on both sides are not the same and thus a current flows through the shielding of the cable, via which the different potentials balance each other out. This equalizing current can be large enough to excessively heat and destroy the shield. However, such a current-carrying shield is worthless; in such a case it is better to ground the cable only on one side.

An example of this is a Pt100 sensor connected via a shielded cable to a controller many meters away, where the shielding is connected to the PE terminal, while the latter is clamped to the sensor at the strain relief and thus connected to the sensor housing. The sensor is located in an electric furnace with a heating capacity of several kW. Since it is quite unlikely that the furnace housing and the switch cabinet are actually at the same potential, the shielding will have little effect. A one-sided grounding at the controller is better in this case because in this example the controller has better grounding. Grounding on both sides is only recommended if the transmission distances are long and it is ensured that the same ground potentials exist on both sides. The following requirements for earthing can be derived from this:

- Ground each device with its own ground wire at a common point;
- Shielded cables must be grounded on one side only, usually on the device;
- The earthing point must have as direct a connection as possible to a potential equalization point (point connected to earth). Do not run earth lines parallel with supply or other current-carrying lines.

In the case of the incorrect earthing of Fig. 3.37, the earthing cables of both devices are brought together and led to the control cabinet door with a screw. The control cabinet is connected to the protective conductor at the terminal strip and the sensor cable is earthed on both sides.

Fig. 3.37 Example of a "wrong" grounding

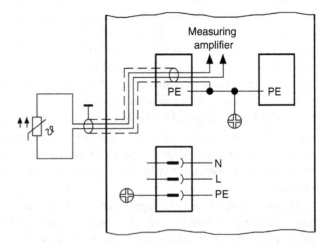

Fig. 3.38 Example of "improved" grounding

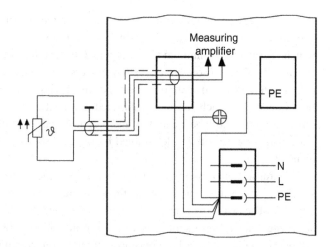

With the improved grounding of Fig. 3.38, the installation has only one grounding point. The control cabinet and its door are earthed with their own wires. The sensor cable is earthed on one side only.

3.4.7 Heat Flow Sensor

By means of a heat flow sensor, the total heat flow emitted or absorbed by a surface can be measured directly. This feature and the easy attachment to the surface to be examined by simply sticking these sensors on opens up a wide range of applications, from the steel producing and steel processing industry to process engineering, heating, and air conditioning technology and aerospace. The heat flow transducers are sensors that are supplied in seven different sizes. Two sizes are rectangular with a width of 12 mm and lengths of 25 or 50 mm, two are square with 25 or 50 mm and three are round with a diameter of 6, 12, or 25 mm. The thickness for all sizes is 2.5 mm. The housing is made of anodized aluminum. They are electrically connected via two copper wires with a length of 30 cm. They are mounted with epoxy resin or adhesive tape. Each element is calibrated and marked with the sensitivity in mV per W/m². These heat flow sensors are divided into three temperature ranges:

- LO series from −20 to +120 °C,
- HI range from −50 to +200 °C,
- BI range from −50 to +100 °C.

The non-linearity is the same for all series and is ±2%. The reproducibility is 0.5%.

The active element of these heat flow sensors is a thermal chain. Its hot compounds lie on the receiving surface and are separated from the cold compounds by a spacer layer. This design ensures optimum sensitivity and independence from the operating temperature. The heat flow sensors have a coating of "HyCalSchwarz", molten colloidal graphite, which guarantees a minimum absorption of $E = 0.9$ under solar radiation.

The heat flow sensors are active transducers and do not require a power supply. They can be interconnected with any writing or displaying a measuring device of sufficiently high impedance. Each sensor is calibrated at a heat flow of 1500 W/m². Because of the direct proportionality between flow and output signal, the calibration curve is fixed over a wide range; the sensitivity determined by the calibration procedure is indicated on each element.

The transducers of the LO series can be used in the range from −20 to +120 °C if high sensitivity is required at the same time. Normally the temperature should not exceed +100 °C. The heat flow can only be measured in one direction. The transducers of the HI series are suitable for a temperature range from −50 to +200 °C. The sensitivity is about 1/8 of the LO series. For these reasons, only the large area types are used here to obtain a larger output signal.

The heat flow sensors of the BI series are very sensitive, therefore they are used only at low temperatures and low heat flows, as well as in cases where heat flow is to be measured in both directions and where it is important that both sides of the sensor have the same sensitivity. The BI series transducers are calibrated on both sides to ensure that both give the same signal.

When installing the LO and HI series, the uncoated side is the transducer. The BI series can be used for bi-directional flows, i.e. when the transducer is installed between two surfaces, and always in applications where the receiving side is on the surface to be investigated. These sensors are attached to the surface with double-sided adhesive tape, epoxy resin, or other adhesive. If an additional thermocouple is installed, the measuring point is marked by a white dot.

3.5   Measuring Mechanical Quantities with Temperature Sensors

Temperature sensors were originally developed only for measuring temperatures. In many cases, however, the advantages of electrical measurements are so considerable that the acquisition of other, e.g. mechanical variables such as level or flow velocity is also useful.

3.5.1   Level Measurement

There are several ways to measure the contents of liquid containers. Commonly used in practice are floats with potentiometers, but this method is mainly suitable for stationary installations, as it requires an extensive linkage for implementation.

In Fig. 3.39 a PTC thermistor is immersed in the liquid. It works with self-heating, i.e. it is heated by its operating current. If it is in the liquid, it dissipates the heat and the

Fig. 3.39 Level control with
PTC thermistor

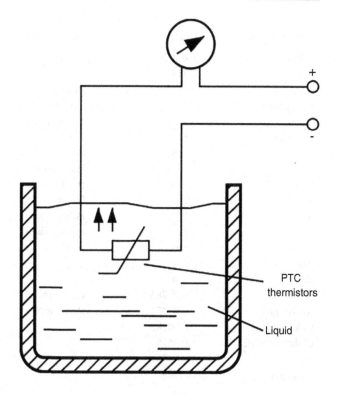

PTC
thermistors

Liquid

resistance is low. A high current flows, which is indicated by the measuring instrument. If the liquid level drops below the sensor, the heat dissipation is omitted and the resistance value increases. This reduces the current.

Only one sensor is shown in Fig. 3.39. In practice, several sensors are often used for a more accurate measurement, at least two, one for the minimum when the vessel is nearly empty and one for the maximum when it is nearly full. In addition, further sensors can also be attached for the intermediate levels.

The thermal transition between the PTC thermistor and the liquid causes a certain time delay, which can cause a measurement error. How quickly the PTC thermistor responds depends primarily on the ratio of the thermal resistance to the heat storage capacity. The greater the thermal resistance, the slower the sensor reacts; this depends on the type of material, and the thickness of the coating. The heat capacity is made up of the specific heat capacity and the sensor mass.

If the response time is to be short, the sensor must be as small as possible and the heat conduction as good as possible. An air gap has a particularly unfavorable effect because air is a very poor conductor of heat. Table 3.4 shows the thermal conductivity λ of various materials.

The thermal conductivity λ of solids is relatively independent of the ambient temperature, whereas that of gases and liquids depends to a greater or lesser extent on the ambient temperature.

Table 3.4 Thermal conductivity λ in at $\dfrac{kJ}{m \cdot h \cdot K}$ the specified temperatures

Substance	λ at 20 °C	Substance	λ at 20 °C
Aluminum	754	Alcohol	0.67
Lead	126	Benzene	0.544
Iron, pure	264	Glycerine	1005
Gold	1118	Transformer oil	0.461
Constantan	81.8	Tuluol	0.544
Copper	1382	Bakelite	0.837
Brass	293–419	Hard tissue	1.26
Nickel	318	Hard paper	1047
Platinum	255	Plexiglas	0.628
Silver	1507	Polyamides	1.26
Steel	126	Pressed fabrics	1.13
Tungsten	603	PVC	0.586
Zinc	419		
Tin	234		
Ammonia	0.00783	Fiber optics	0.1172
Acetylene	0.0737	Rock wool	0.126
Chlorine	0.00285		
Carbon dioxide	0.0515		
Air	0.0875		
Hydrogen	0.443		

3.5.2   Measurement of Flow Velocity

The measuring method of a thermal flow sensor is shown in Fig. 3.40. The NTC1 thermistor is located directly in the flow (gas or liquid), while the NTC2 thermistor is operated as a compensation resistor. Without NTC2, NTC1 reacts not only to the flow velocity but also to the temperature of the medium. NTC1 is a heated element, it is cooled by the flow—the greater the velocity, the stronger. NTC2 is used to measure the temperature of the medium, so it must not be in the flow.

The characteristic of the detection of the flow velocity is the heat transfer between NTC1 and the medium. Again, the smaller NTC1 is, the faster and more accurate the entire device reacts. The separating material between the sensor and the environment must be made of a material with good heat conductivity.

Another type consists of a ceramic substrate with a highly accurate thin-film resistance thermometer and two thick-film heating resistors. Both the temperature sensor and the heating resistors are laser trimmed. The high trimming accuracy ensures the true interchangeability of the sensors and provides consistent measurement sensitivity for each individual specimen.

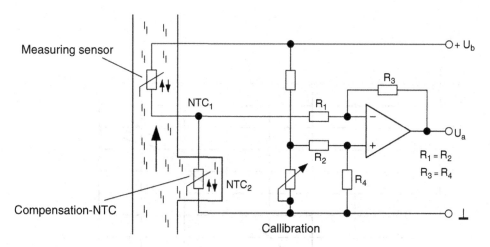

Fig. 3.40 Bridge circuit for recording the flow velocity

These airflow sensors exploit the temperature dependence of a heating resistor in the airflow; they operate in the ambient temperature range from −40 to +70 °C. When voltage is applied, the two heating resistors take up a heating power of approx. 1 W and increase the temperature of the sensor to 90–150 °C. When air flows over it, it is cooled down and changes its resistance. By operating with constant current, this change can easily be converted into a voltage change. These sensors have a non-linear characteristic (Fig. 3.41) with higher sensitivity in the lower part of their measuring range from 0 to 300 m/min.

The resistance characteristics as a function of the airflow are for illustrative purposes only. The actual behavior in the airflow also depends on the thermal conductivity between the sensor and its connector. If these sensors are installed and calibrated, the repeatability of the measured values in practice is ±5%. The application-specific plug connection to the sensor can lead to a loss of sensitivity due to heat transfer and should therefore be thermally insulated as well as possible.

The airflow sensors are particularly suitable for applications where high repeatability, low hysteresis, and high long-term stability are required. However, they react sensitively to temperature fluctuations. They are ideal for applications where the ambient temperature changes very little. For this reason, only versions with temperature compensation are found in practice. Figure 3.42 shows an airflow sensor with temperature detection.

First-order ambient temperature effects are compensated by subtracting the TD sensor output voltage from the actual output voltage of the AW sensor, so that the relative output voltage is limited to changes in the thermal gradient, regardless of changes in ambient temperature.

The internal circuit of Fig. 3.43 shows the separation between the AW sensor with its heater and the TD sensor for temperature detection. The operating circuit is shown in Fig. 3.43. Airflow up to 300 m/min can be detected.

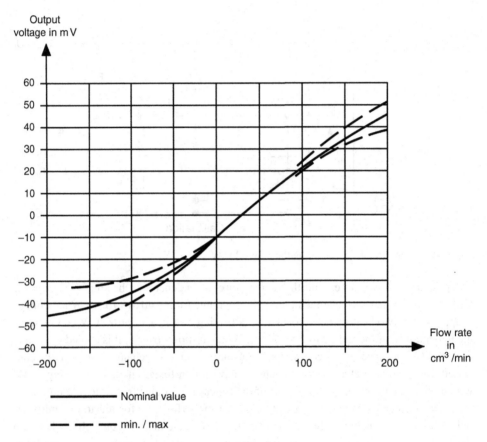

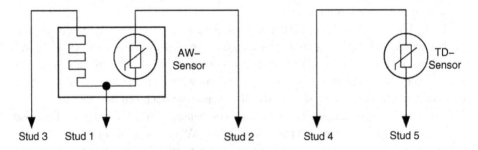

Fig. 3.41 Characteristic curve of an airflow sensor as a function of the output voltage as a function of the flow, referred to the zero voltage

Fig. 3.42 Design of an airflow sensor with temperature measurement

Table 3.5 shows the operating characteristics of the circuit of Fig. 3.43 at an operating voltage with $+U_b = 7$ V.

Fig. 3.43 Generation of the output voltage U_1 of the temperature sensor and U_2 of the airflow sensor

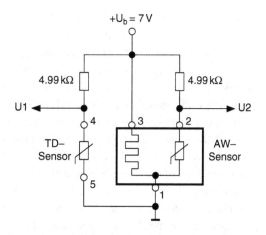

Table 3.5 Example of the voltage curve over the airflow when switching Fig. 3.43

Airflow in m/min	$U_2 - U_1$ for −40 °C	$U_2 - U_1$ for +25 °C	$U_2 - U_1$ for +85 °C
0	0.52	0.49	0.44
5.2	0.48	0.43	0.40
0.5	0.43	0.41	0.37
45.7	0.39	0.37	0.34
61.0	0.37	0.35	0.32
76.2	0.35	0.35	0.31
91.4	0.34	0.33	0.30
106.6	0.33	0.32	0.28

3.5.3 Microbridge Airflow Sensors

While the simple airflow sensors are based on temperature sensors (resistance thermometers), the more expensive and very accurate microbridge airflow sensors work on the principle of the hot film anemometer. The microbridge airflow sensors contain a special silicon chip, which is manufactured according to the latest microstructure technology. It consists of a thermally insulated thin film bridge structure with a heating element and two temperature sensors. The bridge structure ensures high measurement sensitivity and fast response to air or gases flowing over the chip. The two temperature sensors with the heating element in the middle enable the detection of the flow direction and the measurement of the flow rate. Laser-trimmed thick-film and thin-film resistors ensure consistent measurement sensitivity from the sensor to the sensor.

The microbridge air sensor is housed in an elongated housing and has a nipple with a diameter of 5.1 mm at each end. The air system is connected to these nipples. The output voltage ranges between 0 and 45 mV for a volume flow of 0–200 cm³/min, and between 0 and 55 mV for 0–1000 cm³/min. The output voltage can also be calibrated to average duct flow velocity or dynamic differential pressure between the two connection nipples. A specially designed housing accurately directs and controls the airflow over the micro-bridge structure and is easily mounted on a printed circuit board. The data in cm³/min always apply to air at 0 °C and sea level.

The microbridge air sensor uses thermosensitive foils which are applied to a thick film of dielectric material. These are stretched in the form of two bridges over a depression anisotropically etched in the silicon chip. The chip is arranged in a precisely dimensioned airflow channel to ensure a reproducible response to flow. Highly efficient thermal insulation of the heating element and the temperature sensor resistors is achieved by the air space in the etched trough under the airflow sensor bridges. The tiny dimensions and thermal insulation of the micro-bridge airflow sensor are crucial for the fast response and high sensitivity.

The microbridge air sensor is a passive measuring device with two Wheatstone full bridges, one for the closed heating control circuit, and one for the dual-sensing elements. Figures 3.44 and 3.45 show the additional circuitry required for the proper operation of the sensor.

The heating control circuit in Fig. 3.44 ensures that the operating characteristics are maintained. It is specially designed for this sensor and ensures a flow-proportional output and minimization of errors due to fluctuations in the ambient temperature. The circuit causes a constant overtemperature of the heating element above the ambient temperature during changes in temperature and airflow. The ambient temperature is detected by a resistor on the chip similar to the heating resistor.

This type of heating control also reduces the effects of changes in humidity and gas composition but does not completely eliminate them. The changes can affect thermal conductivity and change the operating characteristics of the heating element and the temperature sensor resistance.

The power supply of the measuring bridge shown in Fig. 3.45 is also necessary to maintain the operating characteristics. This is a full-bridge in which the two sensor resistances

Fig. 3.44 Heating control circuit for the microbridge airflow sensor

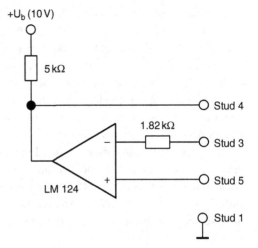

Fig. 3.45 Operating the
measuring bridge for the
microbridge airflow sensor

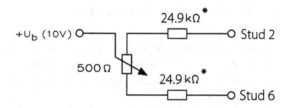

* Resistances between +U_b and stud 2
or stud 6 are balanced with potentiometer

Fig. 3.46 Suitable differential
amplifier circuit for a measuring bridge of the microbridge
airflow sensor

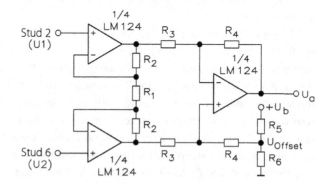

form two active bridge branches. The output voltage of the sensor proportional to the
operating voltage corresponds to the differential voltage on the measuring bridge. When
the flow direction through the sensor is reversed, the polarity of the differential voltage
changes and thus also that of the output voltage of the microbridge airflow sensor.

The differential amplifier circuit in Fig. 3.46 provides a user interface to the measuring
bridge, as it can amplify the sensor signal and shift the zero offset voltage.

In some airflow applications, dust particles may accumulate, but they can be reduced to
a minimum. Larger particles tend to flow in airflow past the microbridges that are parallel
and close to the duct wall. Smaller dust particles are repelled from the heated microbridges
by the "Brownian" molecular movement and strong temperature gradient so that the
microstructure remains clean. Lifetime tests with air currents at 50 cm³/min showed that
the sensor could operate in typical industrial air for 20 years without changing its characteristics. Clogging due to dust adhering to the chip edges and duct walls can be completely
prevented in applications with low air flows by using a simple filter. Since the diameter of
the filter is much larger than that of the duct (3.6 mm), the flow resistance of the filter can
be neglected compared to the duct resistance. This means that even larger accumulation of
dust in the filter has practically no effect on the overall flow resistance. If filtration is
required, a 5 µm filter upstream of the airflow is appropriate.

3.5.4   Hot Film Air Mass Sensor

To measure the mass flow, a resistor is cyclically heated and cooled down. The different temperature increase is a measure for the measured variable. In order to achieve the shortest possible cycle times and high thermal efficiency, the sensor must have an appropriate mechanical design.

Here too, heat is extracted from an internally or externally heated resistor by the medium flowing past. The different temperature increase is a measure of the flow velocity, or more precisely of the mass flow. To determine the actual temperature difference, knowledge of the temperature of the flowing medium is required. Both gases and liquids can be used as heat carriers.

The sensor element of Fig. 3.47 consists of a ceramic substrate, which is equipped with the following thick-film resistors by means of a screen printing process Air temperature sensor resistor $R\vartheta$, heating resistor R_H, sensor resistor R_S, and trimming resistor R_1.

The platinum metal film resistor R_S is kept at a constant excess temperature with respect to the temperature of the outflowing medium by means of the heating resistor R_H. Both resistors are in close thermal contact. The temperature of the incoming air acts on the resistor R_ϑ. The trimming resistor R_1 is connected in series with this, which compensates the temperature response of the bridge circuit over the entire operating temperature range. Together with R_2 and R_ϑ, R_1 forms one branch of the bridge circuit, while the supplementary resistor R_3 and the sensor resistor R_S form the second branch of the bridge. The differential voltage of both resistor branches is tapped at the bridge diagonal as a measurement signal. The evaluation circuit is mounted on a second thick-film substrate. Both hybrids are integrated into the plastic housing of the plug-in sensor.

The hot-film air mass sensor is a thermal flow sensor, the characteristic curve of which is shown in Fig. 3.48. The layer resistances on the ceramic substrate are exposed to the air mass flow to be measured. The sensor is much less sensitive to contamination than, for example, a hot-wire air mass sensor, and a free glowing of the measuring resistors is not necessary.

Fig. 3.47 Circuit of a hot-film air mass sensor

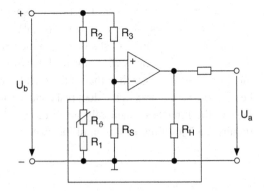

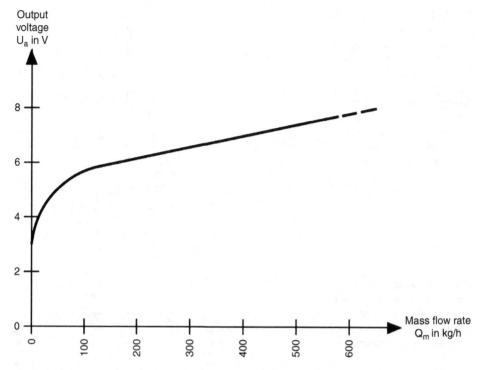

Fig. 3.48 Characteristic curve of the hot-film air mass sensor

When designing the air mass sensor, thermal aspects as well as practical conditions such as handling and prevention of dirt deposits must be taken into account. Figure 3.49 shows the sensor element with its thick-film resistors. No water or other liquid must be allowed to collect in the measuring tube. Therefore it must be inclined at least 5° relative to the horizontal.

3.5.5 Hot-Wire Air Mass Sensor

A hot-wire air mass sensor allows the mass of air or gas flowing through per unit time to be measured independently of density and temperature. The flow diameter for the hot-wire air mass sensor of Fig. 3.50 determines the airflow rate. On the inlet side, a wire mesh protects the hot wire from mechanical impact and homogenizes the flow. On the outlet side, a grid shields the hot wire from backfiring when this sensor is used in automotive applications. The platinum hot-wire trapezoidally stretched in an inner tube, has a diameter of only 70 μm. The attached housing contains the electronic control and cut-off circuit as well as the bridge resistors. The hot-wire air mass sensor operates according to the constant-temperature principle. Here the hot wire is a direct component of a bridge circuit whose output voltage is regulated to zero by changing the heating current. If the air

Fig. 3.49 Typical structure of
a hot-film air mass
sensor:R = air temperature
measuring resistor, R_1 = trimming resistor, R_2, R_3 = supplementary resistors, R_H = heating
resistor (on the back of the
substrate), R_S = sensor resistor
for flow, A = connections of
R_H, S = distance for thermal
decoupling of heater and R_ϑ

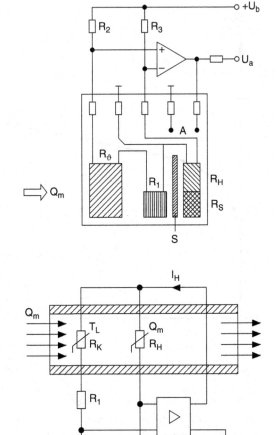

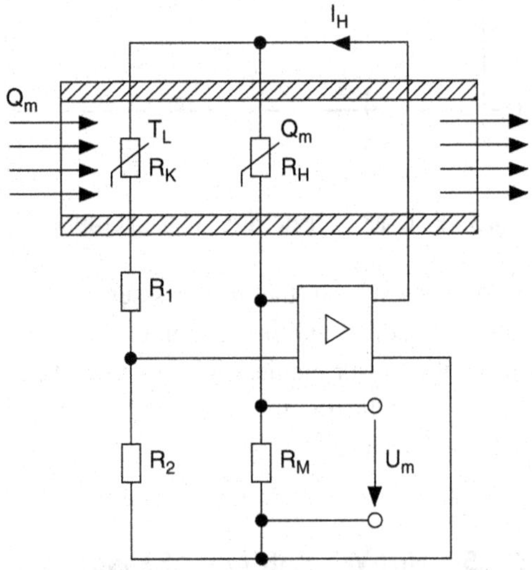

Fig. 3.50 Design of a
hot-wire air mass sensor up to
900 kg/h: R_H = hot-wire
resistance, R_K = resistance of
the temperature compensation
sensor, R_1, R_2 = high-
impedance resistors, R_M = precision measuring resistor,
U_m = signal voltage for air
mass flow rate, Q_m = air mass
per time unit and T_L = air
temperature

volume increases, the wire is cooled down considerably and its resistance decreases. This
leads to a detuning of the resistance ratios in the bridge circuit, which immediately causes
a control electronics to increase the heating current. This increase in current is such that
the hot wire practically maintains its original temperature.

The characteristic curve for the hot-wire air mass sensor is shown in Fig. 3.51. This
means that the required heating current is a measure of the air mass flowing through, independent of the density and temperature of the air. As the sensor has no moving parts, it
works wear-free. Due to the control process, the heat content of the wire does not change,
therefore the sensor reacts very quickly to any change inflow. When the air mass sensor is
operated, deposits can form on the hot wire, which influences the measurement result. It

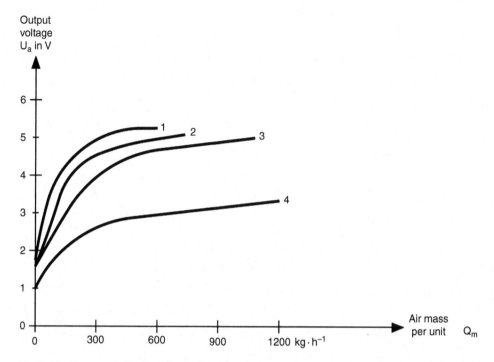

Fig. 3.51   Characteristic curve for the hot-wire air mass sensor

is, therefore, necessary to heat the hot-wire to about 1000 °C for 1 s after a certain period of use in order to burn it free of any dirt. An additional potentiometer is used in automotive applications to adjust the idling mixture.

Optical Sensors

4

Summary

At the beginning of the twentieth century, it was recognized that light is not a continuum, but that it occurs in individual light quanta, the so-called photons. In semiconductors these release electrical charge carriers—electrons and defective electrons (holes). This effect makes it possible to measure the strength of incident light electrically, for example using voltage, current, or changes in resistance. Depending on whether the process of charge release by light takes place on the surface or inside the semiconductor, a distinction is made between the external or internal photo- or photoelectric effect. The optoelectronic components are divided into three main groups:

- Optical-electrical converters or optical detectors: These include photoresistors and photodiodes in various designs, as well as solar cells, photocells, phototransistors, and photothyristors. A distinction is made between passive and active types. In the case of a passive type, only the resistance changes due to the incidence of light, which results in either a current or voltage change. With active components, the generated electrical signal is amplified internally, and a much greater change in current or voltage occurs at the output. Figure 4.1 shows the spectrum of electromagnetic waves. The human eye perceives only a small part of it—visible light. Optical-electrical transducers, on the other hand, are also sensitive to non-visible radiation.
- Electrical-optical converters: Light emitters that convert electric current into light radiation, that is, all types of electric lamps, light-emitting diodes (LEDs), lasers (light amplification by stimulated emission of radiation), picture tubes, displays, etc. Among lasers, semiconductor lasers (laser diodes) are of particular interest.
- Systems that contain both: These include photoelectric barriers and optocouplers. In the case of a light barrier, the transmitter and receiver are optically connected via the light beam. If the beam is interrupted, the receiver registers this and converts it into a

© Springer Fachmedien Wiesbaden GmbH, part of Springer Nature 2022
H. Bernstein, *Measuring Electronics and Sensors*,
https://doi.org/10.1007/978-3-658-35067-3_4

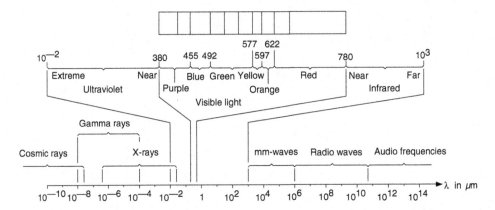

Fig. 4.1 Electromagnetic radiation spectrum with the corresponding applications in technology

suitable output signal. If a light transmitter and a light receiver are combined in one housing, electrically isolated from each other but optically coupled, then an optocoupler is produced. Its main task is the galvanic separation of two circuits.

This chapter also deals with sensors that use light to measure other variables, such as the presence of objects, distances, positions, angles of rotation, etc.

4.1   Properties and Design

Optical sensors take a special position in technology, as they allow a non-reactive determination of the measured variable, that is, this or the measured object is not influenced by the measurement. For many optical measurements, a light emitter (emitter) and a light receiver (detector) must be combined. For this purpose, the optical behavior of these components will be discussed first.

4.1.1   Luminous Sensitivity

In photometry, light is not evaluated according to its energy or power but based on the brightness sensation of the human eye. This is wavelength-dependent. The relative spectral sensitivity or spectral brightness sensitivity $V(\lambda)$ for the human eye is shown in Fig. 4.2.

For the wavelength λ in conjunction with the color tones, see Table 4.1. In the short-wave range, the visible range is followed by ultraviolet ("UV"). This radiation causes strong chemical and biological effects. The likewise invisible range from 780 nm is called infrared or "IR"; it is heat radiation. The human eye has its maximum sensitivity at the wavelength $\lambda = 555$ nm (green). The spectral brightness sensitivity is fixed at $V = 1$.

Fig. 4.2 Spectral brightness sensitivity level for the human eye and the maximum is at a wavelength of λ = 555 nm

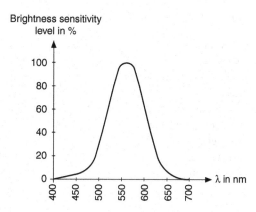

Table 4.1   Wavelength λ with the corresponding color

Wavelength (nm)	Color tone
<360	Ultraviolet
420	Purple
470	Blue
530	Green
570	Yellow
610	Red
>780	Infrared

In addition to the spectral brightness, luminous intensity is an important photometric quantity. It is given in "candela" (cd); 1 cd is the luminous intensity of a radiation source that emits monochromatic light of the frequency of 540 THz of the vacuum wavelength of 555 nm with the radiant intensity 1/683 Steradiant in a certain direction.

The luminance L is defined as the quotient of the luminous intensity and the illuminated area. At a luminance from about L in 75 cd/cm^2, glare occurs in the human eye. Table 4.2 lists the luminances of different light sources.

The luminous flux Φ is defined as the product of luminous intensity and a solid angle radiated through the light. The SI unit of luminous flux is the lumen (lm). Specific luminous emission M is the quotient of luminous flux and radiating area. The quantity of light (Q) defines the product of luminous flux and time, the "lumen second" (1 ms).

Illuminance E is important for the user. It is the quotient of incident luminous flux and the size of the receiver surface. The SI unit for illuminance is the lux (lx). The illuminance decreases with the square of the distance.

$$E = \frac{\Phi}{A} \text{ in lx} = \frac{\text{lm}}{\text{m}^2}$$

Table 4.2 Luminance of different light sources

Light sources	L in cd/cm^2
Night sky	0.001
Night sky with moonlight	0.25
Candlelight	1
Frosted tungsten lamp	40
Clear tungsten lamp	1000
Daylight	5000
Midday sun	50,000
Glacier sun	150,000

Table 4.3 Illuminance for some practical applications

Illuminance	E in lux
Moonless clear night	0.005
Night of a full moon	0.2
Lighting in living spaces	100
Lighting in offices	150
Lighting in schools	200
Lighting at workplaces	300
Overcast winter sky	500
Overcast summer sky	5000
Sunlight in winter	20,000
Sunlight in summer	100,000

The luminous intensity I can be determined from the total current using the light distribution curves. Table 4.3 shows the illuminance for some applications.

Exposure H is defined as the product of illuminance and time. The SI unit for exposure is the lux second (lx s). This unit is familiar with the light meter in photography, where the required exposure time for the film is determined from the illuminance.

4.1.2 Photodiode Technology

Photodiodes are manufactured using the planar process. The edges of the PN junction lie protected under the SiO$_2$ used as a diffusion mask, which was created by oxidation of the silicon surface. Due to the low dark current, photodiodes are very well suited for the detection of low light signals and operation at high reverse voltages.

A special type of photodiode is the PIN photodiode. Here, there is a large, high-impedance, intrinsic I-zone between the P- and N-zone. The main advantages of PIN photodiodes are extremely short switching times combined with high infrared sensitivity. Thanks to specific technological measures, they get by with a low operating voltage.

Diodes with a large space-charge width are called PIN diodes, regardless of whether an originally intrinsically conductive (I) crystal has been P- or N-doped on the opposite surfaces or whether diffusion takes place here into a very high-impedance, low-doped substrate material so that the space-charge width becomes large in this way. In the drift field of the space-charge zone, the charge carriers generated are collected within short times (ns range). However, even in the low-frequency range, e.g., for infrared sound transmission or infrared remote control, the use of PIN diodes offers considerable advantages. Relatively large-area diodes with very low capacitance can be produced. Consequently, the diodes can be operated at a low operating voltage and high load resistances (e.g., 100 kΩ), resulting in high signal voltage levels.

Another important design is the avalanche or avalanche photodiode. It is particularly well suited for detecting modulated radiation at low signal levels, high bandwidths, and small light-sensitive areas. The internal amplification of the photocurrent is achieved by a multiplication process in the high field of the space-charge zone of a reverse-polarized PN junction. The diodes are operated below the breakdown voltage. The internal amplification M—the ratio of the photocurrent I_{Ph} during reverse voltage operation to the photocurrent at a low reverse voltage (approx. 5 V)—can be determined by the applied reverse voltage and reaches values of up to more than 200 for silicon versions.

In the frequency range between 10 and 100 MHz, the optimal multiplication factor for the detection sensitivity is around 50. Avalanche diodes are operated in current mode up to frequencies of 50 MHz with broadband amplifiers. At microwave frequencies in the GHz range, avalanche diodes can be operated with low load resistances (<100 Ω) and corresponding voltage amplifiers.

From frequencies of 1 MHz on, the photo-avalanche diodes are generally preferred to PIN diodes, whereby aspects such as the complexity of the preamplifier, optical adjustment possibilities, and a must also be taken into account.

At higher frequencies, the PIN diodes limit the thermal noise of the load resistor or the preamplifier for the detection sensitivity. However, the internal amplification of avalanche diodes allows the photo signal to be boosted above the noise of the load resistor. As a result, avalanche diodes are superior to PIN diodes at higher frequencies. Avalanche diodes are therefore ideally suited for the technique of transmitting information via fiber optics and e.g., for distance measurement.

4.1.3 Applications of Photodiodes

The most important measurements on a photodiode are dark and light measurements, both are important for use as sensors. Additionally, the open-circuit voltage U_0 and the short-circuit current I_K can be measured.

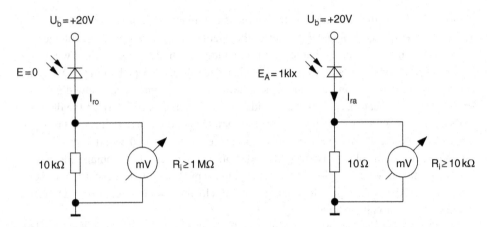

Fig. 4.3 Measuring circuit for the dark blocking current (*left*) and the light-blocking current of photodiodes (*right*)

The dark blocking current I_{ro} must be measured in absolute darkness since silicon photodiodes have blocking currents in the nA range and illuminance of a few lux is already sufficient to falsify the measured value. If a high-impedance digital voltmeter is used for the measurement, a measuring resistor is connected in series to the test object. This resistor must be dimensioned in such a way that the voltage drop occurring on it remains small compared to the operating voltage. The change in the reverse voltage at the measured object can then be neglected. Figure 4.3 (left) shows the circuit for measuring the dark blocking current. The light-blocking current of photodiodes is measured like the dark blocking current, but the photodiode is now irradiated. The measuring resistor must be low impedance because of the higher currents.

A calibrated, unfiltered tungsten incandescent lamp serves as the light source for the bright measurements. The lamp current is set to the color table 2855.6 K, the standard illuminant A according to DIN5033. The prescribed illuminance E (usually 100 lx or 1000 lx) is achieved with the aid of an optical bench by changing the distance α between the lamp and the photodiode. It can be measured with a $V(\lambda)$-corrected lux meter or, if the luminous intensity I of the lamp is known, according to the relationship

$$E_v = \frac{I_v}{\alpha^2}$$

calculate. This so-called "photometric distance law" applies to point light sources, that is, with dimensions of the light source (i.e., the filament) that are small ($\leq 10\%$) compared to the distance α to the receiver component.

A photodiode is suitable for the visible and near-infrared radiation ranges and can therefore be used in light barriers, optical scanning, and counting devices or in data transmission. If a silicon avalanche photodiode is inserted at the end of an optical fiber cable, a broadband detector for demodulating fast signals is obtained. Figure 4.4 shows an optical

Fig. 4.4 Optical receiver for a
fiber optic cable

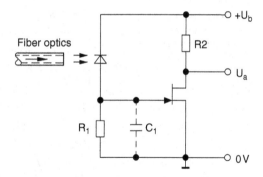

receiver for FO signals with photodiode and amplifier. The N-channel FET works in the
source circuit, that is, the input signal is inverted for the output signal.

The photodiode is located in a voltage divider that drives the gate of the FET. The input
capacitance C_1 (dashed line), which limits the frequency response of the amplifier, must
also be taken into account. The cut-off frequency f_g can be calculated using the formula

$$f_g = \frac{1}{2 \cdot \pi \cdot R_1 \cdot C_1}$$

and the resulting values are from 10 to 500 MHz.

4.1.4 Photoresistance

The photoresistor (LDR, Light Dependent Resistor) is a passive component that changes
its resistance depending on the illuminance. The effect is based on the release of charge
carriers by the light. The semiconductors used are cadmium sulfide (CdS), cadmium sel-
enide (CdSe), cadmium telluride (CdTe), and some lead and indium compounds.
Depending on the material, the spectral sensitivities vary greatly (Fig. 4.5). The CdS type
is mainly used for photographic purposes, as the spectral sensitivity curve corresponds
almost exactly to that of the human eye. Figure 4.6 shows the characteristic curve of the
photoresistor LDR03 and the circuit symbol.

The characteristic curve of Fig. 4.6 shows the very high light sensitivity of a photoresistor. The dark resistance is very high and lies between 1 and 10 MΩ depending on the type.
The light resistance, on the other hand, takes on values from 10 Ω to 1 kΩ. To achieve the
highest possible sensitivity, if the resistance of the photoresistor is chosen accordingly
high, the photosensitive layer must be arranged in a meandering pattern.

A photoresistor has no barrier layer. For this reason, the generated charge carriers need
a relatively long time to recombine. Photoresistors, therefore, react very slowly, their cut-
off frequencies are in the range from 20 to 500 Hz. In contrast to other optoelectronic
components, they can be operated with direct or alternating current. The temperature coefficient is low at less than 1%/K, but the maximum permissible operating temperature

Fig. 4.5 Spectral sensitivity of photoresistors made of different semiconductors

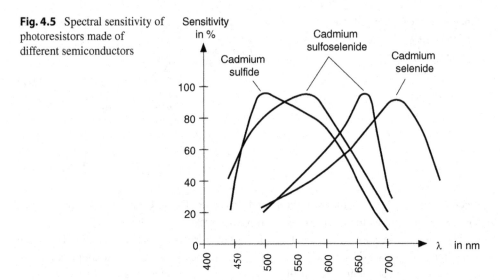

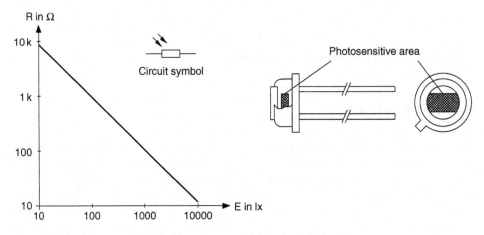

Fig. 4.6 Characteristic curve of the photoresistor LDR03 and designs

should not exceed 70 °C. The power dissipation at an ambient temperature of 40 °C should generally not be greater than approximately 100 mW. The maximum permissible operating voltage is between 50 and 350 V depending on the type.

4.1.5   Measuring Circuit with a Photoresistor

The simplest measuring circuit with a photoresistor is shown in Fig. 4.7. It is connected in series with an ammeter and a battery. The incident light causes the resistance to change and the current to adjust accordingly. This rather simple circuit can hardly be used in

Fig. 4.7 Exposure meter or lux meter

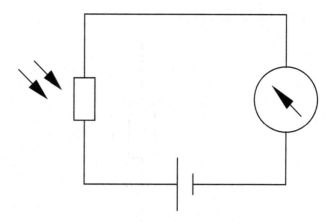

combination with integrated circuits and microprocessors or microcontrollers. It is better to use the voltage divider shown in Fig. 4.8. From the characteristic curve of a photoresistor, one gets e.g., the values 900 Ω for 100 lx and 100 Ω for 1000 lx. The output voltage is calculated from

$$U_a = U_e \cdot \frac{R_{Ph}}{R_1 + R_{Ph}}$$

$$U_{a100} = 5\,\text{V} \cdot \frac{900\,\Omega}{1k\Omega + 900\,\Omega} = 2.36\,\text{V}$$

$$U_{a1000} = 0.45\,\text{V}$$

The output voltage, therefore, fluctuates depending on the brightness between 2.36 V at 100 lx and 0.45 V at 1000 lx. This voltage can be used to drive an AD converter, which then generates digital information from the analog value.

The turbidity of liquids can be measured with a photoresistor. The liquid to be examined is located in a transparent container or tube. On one side there is a lamp, on the other side the photoresistor. The light transmission can be used as a measure of the turbidity of the liquid because the light beam must penetrate this medium, whereby its intensity decreases.

4.1.6   Twilight Switch

With this circuit, switching operations can be triggered automatically at nightfall, e.g., yard, house, garden, house number, or path lighting can be switched on. The switch-on threshold can be infinitely varied using an adjuster. However, the circuit is only suitable for incandescent lamps, not for energy-saving lamps or fluorescent tubes.

Figure 4.9 shows the photoresistor as a light sensor, the amplifier section for evaluation and control of the thyristor, the mains section (230 V) with the thyristor, and the bulb.

Fig. 4.8 Voltage divider with a photoresistor

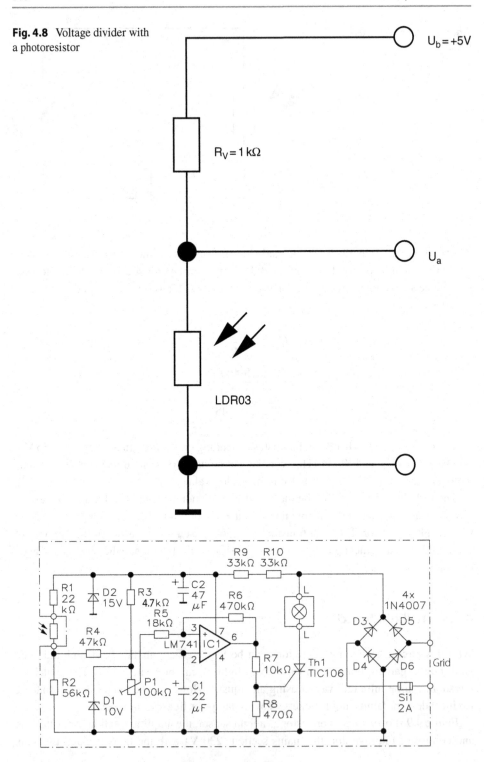

Fig. 4.9 Circuit of a twilight switch for incandescent lamps

Thyristors and TRIACs can be used as contactless semiconductor switches. Instead of a relay, the twilight switch has a thyristor that switches the incandescent lamp on and off. If you use a relay instead of the thyristor, you have to modify the circuit a little bit, but then you can also control energy-saving lamps and fluorescent tubes. When switching such light sources with a thyristor, the thyristor can be destroyed.

A thyristor allows the current to pass only in one direction, as with a diode. Here it is therefore operated in the load circuit with the lamp behind a bridge rectifier. This gives it a pulsating DC voltage with which it can operate. The load connected in series with the thyristor, that is, the bulb, is always switched on when the gate receives current from the electronic circuit. At the end of each half-wave, that is, all 10 ms, the thyristor is automatically extinguished and must be re-ignited by the control electronics.

To select the thyristor, the RMS value must be converted to the peak value, which gives the value $\sqrt{2} \cdot 230\,\text{V} = 325\,\text{V}$. The thyristor must be able to withstand this voltage in the blocked state without a breakdown. The used type TIC106 is suitable for voltages up to 400 V.

Since the circuit is operated directly from the mains, no separate power supply is required. The operating voltage for the operational amplifier is obtained directly from the bridge rectifier via the two 33-kΩ resistors. This pulsating DC voltage is smoothed by the capacitor C_2 and stabilized by the Z-diode at 15 V. This voltage stabilization is sufficient for safe operation, as the power consumption of the components used is very low. For reasons of better heat dissipation, two series resistors R_9 and R_{10} are used in the circuit.

Two circuit branches are connected to the two inputs of the operational amplifier. The circuit branch at the inverting input consists of a voltage divider with the photoresistor. If the incidence of light at the photoresistor increases, its resistance decreases, and the voltage at the inverting input increases. If, on the other hand, the incidence of light decreases, the resistance increases, and the voltage decreases. If a brief fluctuation in brightness occurs—for example, from a large rain cloud or a bird flying past—this is compensated for by capacitor C_1.

While the photoresistor generates the actual value for the twilight switch, the setpoint for the adjuster P_1 can be set over a wide range. To keep voltage fluctuations of the mains away, this voltage divider also has a separate Z-diode. The partial voltage tapped at adjuster P_1 $(0 \ldots 10\,\text{V})$ is thus decoupled from all conceivable interference influences, which has a very positive effect on the stability of the circuit.

The operational amplifier is operated as a Schmitt trigger. The size of the hysteresis can be determined by the resistors R_5 and R_6. Due to positive feedback, the output voltage of the operational amplifier is switched either to ≈ 0 V or to approx. +14 V. In the former case, no current can flow into the thyristor gate, which is blocked and the lamp remains dark. If, on the other hand, the output switches to $U_a = +14$ V, a gate current flows; the thyristor ignites and the lamp lights up.

Due to the photoresistor, there is a voltage of +4 V at the inverting input and +5 V at the non-inverting input. The output voltage of the operational amplifier is therefore in negative saturation. If the brightness at the photoresistor increases, the voltage rises from +4 to

+5 V, for example. If the voltage exceeds the value of +5 V, the output of the operational amplifier flips to the positive saturation voltage of $U_a = +14$ V and a gate current flows into the thyristor. If the brightness at the photoresistor decreases, the voltage at the inverting input decreases; when a certain threshold is reached, the output of the operational amplifier flips back to negative saturation.

The circuit may only be put into operation if it is safe to touch and is accommodated in housing in compliance with VDE regulations.

4.1.7   Phototransistor

In the phototransistor, the photocurrent generated at the collector-base diode is increased by the current gain β of the transistor; typical values are between 100 and 500. This means that in many applications one amplifier stage can be saved. The essential characteristics of a phototransistor can be read from its equivalent circuit diagram, in which the (usually large-area) collector-base diode is located as a photodiode in the input of a normal NPN transistor, which operates in an emitter circuit.

The PN junction of the base-emitter diode in a phototransistor serves as a photoelement. The light falls on the transistor through a small glass lens that is fused into the housing. The resulting photovoltage is effective as base-emitter voltage and controls the transistor accordingly (Fig. 4.10). Since the collector current is greater than the base-emitter current by the current amplification factor, an amplification effect is added here.

In the simplest case, a phototransistor has only two connections: emitter and collector. An externally supplied base-emitter voltage is not necessary for operation, as this is generated in the transistor itself when light falls on it. A phototransistor without base connection is also called a "photoduodiode". However, there are also versions with a base connection, in which the cut-off frequency can be slightly increased by applying a base-emitter voltage. Besides, the operating point of the circuit can be influenced, but this leads to a reduction of the photosensitivity. The characteristic curve $I_C = f (U_{CE})$ of a phototransistor (Fig. 4.11) corresponds to that of normal transistors, the only difference being that the illuminance in lx is entered as a parameter at the point of the base current $_{IB}$.

Fig. 4.10 Operation and circuit symbol of the phototransistor

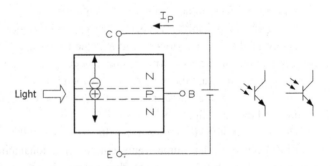

Fig. 4.11 Characteristic diagram of a phototransistor

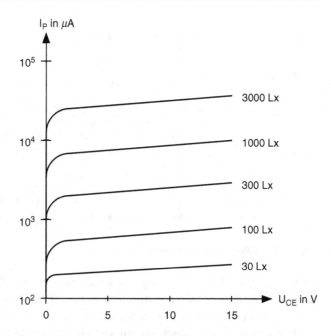

If a particularly high amplification is required, it is recommended to use a photo Darlington transistor which contains two internal amplifier stages in cascade connection. However, this has a lower cut-off frequency.

By optimizing standard processes and by applying new processes in the production of phototransistors, improvements have been achieved in recent years, especially in the following areas:

- Sensitivity in defined spectral ranges
- The linearity of the photocurrent characteristic curve
- Response times: for photodiodes in the nanosecond range, for phototransistors in the microsecond range
- Stability

4.1.8 Automatic Garage Lighting with a Phototransistor

A practical application for a phototransistor is automatic garage lighting. When you come home in the evening and have parked your car in the garage, you laboriously fumble to the light switch. This lighting control should prevent this. When you enter the garage, you flash the headlamp flasher briefly and the garage lighting is immediately switched on for a pre selectable time (up to about 3 min).

The circuit of Fig. 4.12 works with an operational amplifier 741 and a timer module 555. The light signals received by the phototransistor (from the high beam of the car) are

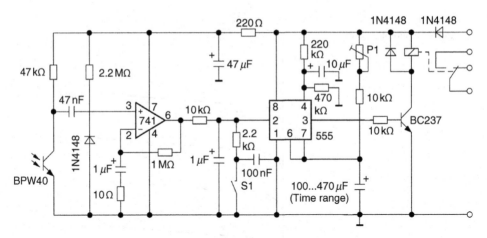

Fig. 4.12 Automatic garage lighting circuit

converted by the 741 into "0" signals, which triggers the 555 and switches on the relay. Slow light changes on the phototransistor do not affect. With the adjuster P_1, the duration can be varied continuously from approx. 2 s to 3 min.

With a separate light switch, the garage lighting can also be switched on for the time set with P_1 when entering the garage. If continuous light is desired, the light switch S_1 remains switched on, whereby after switching off the light remains on for the set time and then goes out automatically. The phototransistor does not need to be shielded from the ambient light.

The relay for switching the mains voltage is located directly on the board. The circuit is operated from a small power supply unit with $U_b = 12$ V to $U_b = 15$ V.

The circuit board is mounted on the back wall of the garage where it is illuminated by the headlights when the car enters. The height can be easily determined by placing the car in front of the wall approximately 2 m and switching on the high beam. The housing with the phototransistor is now mounted where the light cone shines brightest on the garage wall. When driving into the garage with a low beam, a short flash with the headlight flasher is sufficient.

4.1.9 Photoelectric Cell

Photoelements or solar cells are constructed similarly to photodiodes. While with conventional PN-junctions in diodes or transistors the diffusion voltage generated inside cannot be loaded, this is possible with photo elements because of the incidence of light. This generates new pairs of charge carriers at the boundary layer again and again—the greater the intensity, the more. Figure 4.13 shows the mode of operation and the circuit symbol.

Photoelements manufactured according to the mesa method show relatively high leakage currents due to the open PN junction, that is, a low internal resistance at low

Fig. 4.13 Mode of operation and circuit symbol of the photoelectric cell

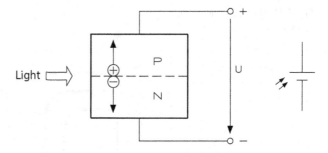

Fig. 4.14 Characteristic curve $U_0 = f(E)$ of a silicon photoelement

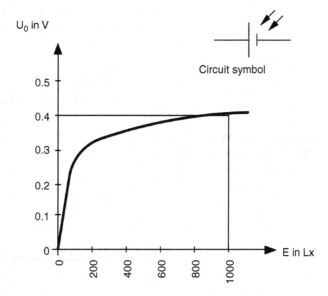

illumination. Due to their low junction voltage, they are particularly suitable for photovoltaic operation. Their advantages are the high light sensitivity and low effort in the production of large-area structures.

While with conventional PN junctions, the diffusion voltage present inside cannot be loaded, this is possible when light is incident on photovoltaic elements and solar cells. This is the case because in the boundary layer of the PN junction charge carrier pairs are repeatedly created by the incidence of light. The PN junction of solar cells is therefore designed over a large area with meander-shaped connections. Figure 4.14 shows the open-circuit voltage U_0 as a function of the illuminance E of a silicon photoelement.

If a photo element or a solar cell is loaded in a circuit, the voltage drops according to a load characteristic curve shown in Fig. 4.15.

Figure 4.16 shows the characteristic curve field of a light-dependent PN-transition, which shows the relationship between the photodiode and a solar cell as well as a normal silicon diode.

Figure 4.17 gives an example of how the luminous intensity of a light-emitting diode can be determined with a photoelement. The measured luminous intensity I_e is multiplied

Fig. 4.15 Load characteristic of a silicon photoelement

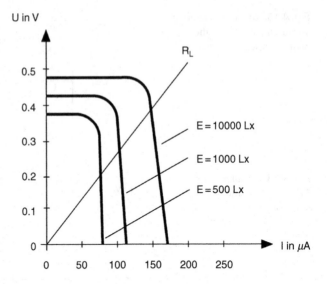

Fig. 4.16 Characteristic diagram of a light-dependent PN transition

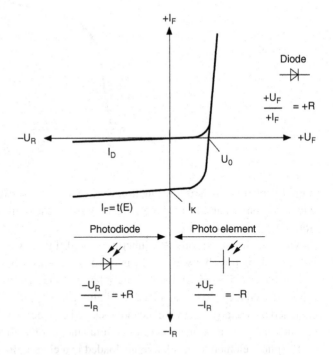

by the absolute eye sensitivity $K_m \cdot V_\lambda$. The wavelength of the emitted radiation of the measured object must be known very precisely. For series measurements, therefore, a calibrated silicon photoelement with a special color filter attached is used to simulate the red edge of the eye sensitivity curve. The BPW20 sensor cell can be used as the photo element, which has a linear short-circuit current characteristic even at the lowest irradiance levels.

Fig. 4.17 Measuring circuit
for a photovoltaic cell

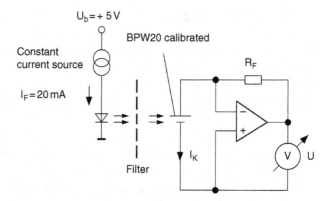

Because of the lower radiation flux of light-emitting diodes compared to IR diodes and because of the color filter, the photo element emits only a few nano-amperes. Therefore an amplifier with a very high impedance input is required.

The voltage of the photoelement can also be measured directly with a high-impedance voltmeter. Photoelements have a characteristic curve of sensitivity as a function of the wavelength of light. Depending on the type, the maximum is more in the blue or red range.

Light comparison circuits can be used in production to measure the differences in brightness or color of mass products, e.g., to control paints or monitor chemical solutions. For this purpose, two similar photoelements and amplifiers are used, usually in a bridge circuit. With one branch a normal pattern is scanned, with the other branch the target is scanned. The difference is amplified, displayed, or fed via an AD converter to a measuring system with a microprocessor or microcontroller.

4.2   Active Optoelectronics

The effect of the active components of optoelectronics is based on the barrier-layer photo-electric effect. Charge carrier pairs are generated by the absorption of light in the semiconductor material. The minority charge carriers are collected at the PN junction, causing a photocurrent to flow in the external circuit.

The usual division of optoelectronic components into the emitter, detector, and coupling elements inevitably also results when describing the manufacturing processes. Emitter components consist in the context valid here exclusively of III-V compound semiconductors such as GaAs (gallium arsenide), GaAsP (gallium arsenide-phosphide), GaP (gallium phosphide), and similar. In contrast, receiver components for visible radiation and short-wave IR radiation are usually silicon components whose technology is similar to that of standard diodes or transistors. In the case of optoelectronic coupling elements, the know-how is mainly in housing and construction. Here, the attempt is made to produce a compact component by skillful adaptation of emitter and detector via a suitable coupling medium.

4.2.1   Emitter Components

The wavelength of the radiation emitted by luminescent diodes is determined primarily by the semiconductor material used and secondarily by its doping. GaAs diodes emit in the infrared range between 800 and 1000 nm. There are essentially two manufacturing processes, which differ mainly in the production of the PN junction:

- To form the PN junction, zinc is diffused into single-crystal N-doped GaAs wafers either over the entire surface, whereby the elements subsequently produced from the wafer by cutting up have a PN junction extending to the open edge (mesa technique), or through photolithographically produced windows in suitable masking layers on the surface (planar technique).
- On single-crystal N-doped GaAs wafers, a thin single-crystal GaAs layer is deposited from a silicon-doped melt by a liquid phase epitaxy process, whereby the PN transition is created by the different incorporation of the silicon into the GaAs crystal lattice at the beginning and towards the end of the process.

Zn-doted IR diodes have shorter response times (1…100 ns) and a comparatively smaller radiation flux (0.5…2 mW), while Si-doped diodes achieve an optical output power of up to approx. 20 mW with response times of several hundred ns.

4.2.2   Laser Diodes (Semiconductor Lasers)

The GaAlAs double hetero laser consists of layers epitaxially deposited from the liquid phase. In most cases, four to five successive layers on a GaAs substrate are used for the continuous wave laser. The P-GaAs layer is the area responsible for the emission. The properties and thickness of this layer (less than 1 μm) must be controlled very precisely during production. Usually, the structure of the stripe laser is used. The lateral confinement of the active area, which is several hundred micrometers long, can be done by different methods, e.g., by mesa etching or by proton implantation. The width of the active area is smaller than 20 μm.

The laser chip is mounted on the best possible heat sink, for which, for example, a special diamond is used. The fact that the radiating area has a size of only 1–20 μm^2 results in very high radiance; typical values are greater than 200 kW/(sr cm^2) (sr $\triangleq$ Steradiant).

Semiconductor lasers or laser diodes (LD) have several advantages over luminescent diodes (LEDs in the IR range IRED), which are mainly used in long-range technology via fiber optic cable, even at longer wavelengths. They generate coherent optical radiation with a very narrow emission width and are sometimes even limited to a single wavelength (monochromatic LD). Their radiation is highly bundled and can be coupled up to 50% into a single-mode fiber, resulting in optical transmission power of several mW in the bevel. This corresponds to an optical transmission level of about −1 dBm. For the wavelengths,

1300 and 1550 nm laser diodes in heterostructure of III/V solid solutions (InGaAsP) are used without exception.

The name laser comes from the English term: Light amplification by stimulated emission of radiation and means something like Light amplification by stimulated emission of radiation.

The basic structure of an LD is similar to that of an edge-emitting LED.

To achieve a light-enhancing effect in the crystal, it is mirrored on both sides so that the existing light is constantly reflected back and forth, thus adding new light to the existing light. As a result of interference from the light waves reflected back and forth, only a few specific wavelengths or, in the case of corresponding resonators, even only a single wavelength can propagate. The mirrors are made semi-transparent. This allows some of the optical radiation to exit the crystal in a highly concentrated form. Figure 4.18 shows the principle of a semiconductor laser or laser diode. Figure 4.19, on the other hand, shows the conditions for wave guidance in the crystal to achieve the bundling and the single wave radiation that is as single wavelength as possible.

The precisely manufactured crystal faces are used as the outer mirror surfaces of the semiconductor laser (mirror resonators of the Fabry-Pérot type). The length of the crystal thus determines the longitudinal restriction of the light rays capable of propagation in the crystal, which are generated as stimulated emission by the injection current at the PN junction. The transverse (vertical) confinement is determined by the thickness of the active layer which is generally 0.1–0.2 μm. The doping and the mixture of materials in the active layer must be designed in such a way that the two adjacent layers of P and N material are completely passive to optical wave propagation despite their respective doping. There are essentially two methods to achieve the lateral (lateral) confinement of the optical waveguide inside the crystal:

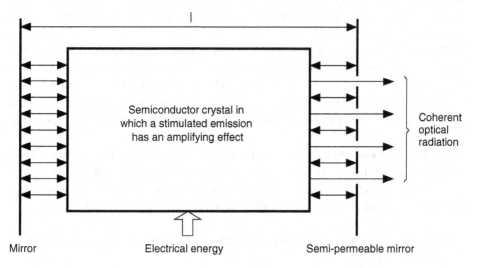

Semiconductor crystal in which a stimulated emission has an amplifying effect

Coherent optical radiation

Mirror   Electrical energy   Semi-permeable mirror

Fig. 4.18 Principle of a semiconductor laser or laser diode

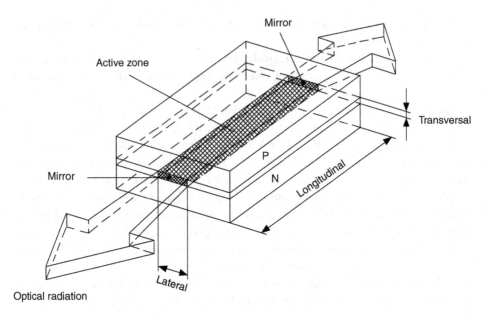

Fig. 4.19 Optical waveguide conditions in a semiconductor laser crystal

- Current-induced lateral wave guidance (gain-guiding—gg) using an oxide strip construction
- Index-guiding (ig) lateral restriction of the wave propagation by a built-in waveguide

An oxide stripe laser as gg-LD uses a stripe contact for the lateral wave guidance, which is caused by the corresponding mask formation of the oxide layer on the top side of the crystal. The geometry of the P-contact thus causes current-induced lateral guidance. Figure 4.20 shows an oxide stripe laser for the wavelength 1300 nm made of InGaAsP on InP substrate.

The depicted InGaAsP-gg laser for the central emission wavelength of 1300 nm emits an optical radiation power of 6 mW for the fundamental mode at an injection current of $I_F = 150$ mA. Its spectral width is $\Delta\lambda = 10$ nm with only four minor modes. It can be modulated up to 500 Mbit/s. The coupling efficiency even for single-mode fibers is 50%, so that up to 3 mW optical radiant power, which corresponds to an optical level of −1 dBm, can be coupled in.

A much better waveguide structure in the crystal compared to gg lasers is obtained with index-guided laser diodes, also known as ig lasers. To achieve the built-in lateral waveguide by a lateral refractive index difference, a second epitaxy process is necessary after the first epitaxy and a corresponding etching process. There are two possibilities for this:

- Liquid phase epitaxy as a conventional method (LPE method—Liquid phase epitaxy) with high failure rates

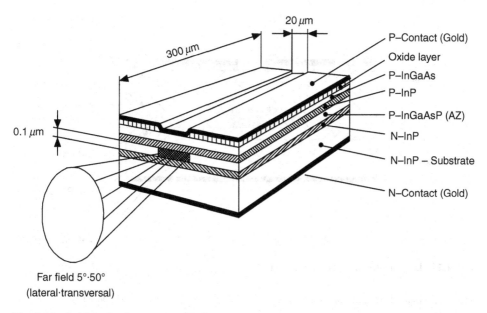

Far field 5°·50°
(lateral·transversal)

Fig. 4.20   Oxide stripe laser made of InGaAsP as gg-LD

- Vapor phase epitaxy (VPE) as a more cost-effective method suitable for larger crystals.

In the XLPE process, two designs are used in practice:

- PBRS-Laser (PBRS—planar buried ridge structure)
- MT-BH Laser (MT-BH—mass transport in buried heterostructure)

While in the PBRS laser crystal the active zone (AZ) is surrounded, that is, also later-ally, by InP with the help of etching after first epitaxy (LPE method) by second epitaxy (VPE method), the MT-BH laser can almost completely do without second epitaxy. Still of the first epitaxy (LPE process), the upper N+-InP layer is under etched mushroom-shaped and the MT layer is filled for lateral wave guidance by a mass transport medium in form of a halogen gas through InP, which is also transported. Figures 4.21 and 4.22 show these two laser constructions.

For both laser designs shown, operationally reliable and durable LDs for the 1300-nm and 1500-nm ranges for threshold currents of only 15 mA are available at relatively low cost, which can be modulated up to 3 GHz. By special longitudinal structuring measures in the form of a wavelength-adapted grating structure above or below the active zone, side modes can be completely avoided for both types so that only a single wavelength is emit-ted. These laser diodes are called DFB-BH lasers (DFB—distributed feedback). With such LDs, bit rates up to 8 Gbit/s can be transmitted over long distances from 100 km with single-mode fibers in the 1500 nm range.

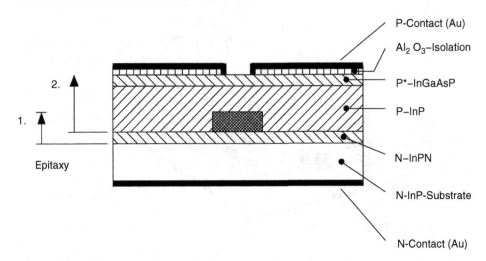

Fig. 4.21 Lateral structure of a PBRS laser

Fig. 4.22 Lateral structure of a MT-BH laser

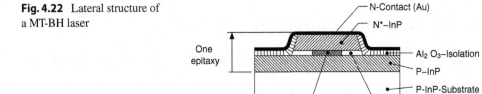

The light amplifying effect of a laser diode starts above a certain threshold current $_{IS}$. This threshold current is strongly dependent on temperature and age. For this reason, laser diodes must be operated with a complex and special temperature control and special control electronics for aging. Figure 4.23 shows the current-power characteristic of an ig-laser and Fig. 4.24 its spectral emission width above the threshold current I_S, whereby the aging phenomena and the temperature response are indicated in Fig. 4.23.

4.2.3   Light-Emitting Diodes

Luminescent or light-emitting diodes (LEDs) emit optical radiation by spontaneous recombination in the active zone of a PN junction operated in the forward direction. In the visible spectral range between blue (around 450 nm) and red (around 650 nm), the materials of GaN (blue), GaP (green), and GaAs with additives such as phosphorus (P) as GaAsP (red) are suitable for the manufacture of LEDs in a wide variety of designs for every

Fig. 4.23 Current performance characteristic of
an ig-LD

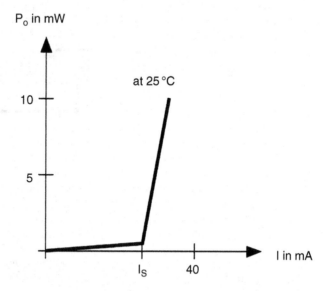

Optical
radiant
power

P_o in mW

at 25 °C

10

5

I_S 40

I in mA

Fig. 4.24 Spectral emission
width of an ig-LD

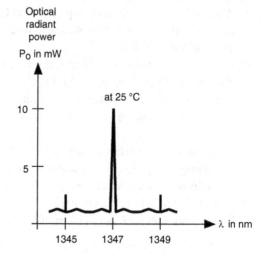

Optical
radiant
power

P_o in mW

at 25 °C

10

5

λ in nm

1345   1347   1349

application of display elements. The basic structure of an LED and its designs are shown
in Fig. 4.25.

LEDs for the visible part of the spectrum are made of GaAsP or GaP. For all colors the
advanced planar technology with covered PN-junctions is used, which provides a long
lifetime. Material production, on the other hand, uses two different technologies:

Fig. 4.25 Basic structure of
an LED and the commercially
available designs

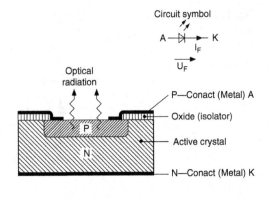

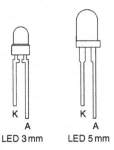

K   K
A   A
LED 3 mm   LED 5 mm

- Red ($GaAs_{0.6}P_{0.4}$): Here an N-type epitaxial GaAsP layer is deposited on a single crystalline GaAs substrate. The phosphorous content is continuously increased with the layer thickness to 40%.
- Green, yellow, and orange: These epitaxial layers are produced in the same process. The substrate here is monocrystalline GaP, which is transparent to the emitting radiation. With a reflective backside metallization, the efficiency can be doubled because no light is absorbed in the substrate.

A total of three materials are available for these colors. Common to all these technologies is nitrogen doping, which increases the light yield of these materials enormously. Figure 4.26 shows the operation of a light-emitting diode with a series resistor R_v. This is always necessary to limit the current I_F.

The forward voltage U_F of light-emitting diodes is largely dependent on the material:

$$
\begin{aligned}
&\text{GaAs - IR - Dioden}: && U_F \approx 1.2\,\text{V} \\
&\text{Rote GaAsP - LEDs}: && U_F \approx 1.6\,\text{V} \\
&\text{Grüne GaP - LEDs}: && U_F \approx 1.8\,\text{V} \quad \text{green, red, blue} \\
&\text{Blaue GaN - LEDs}: && U_F \approx 2.4\,\text{V}
\end{aligned}
$$

Fig. 4.26 Operation of a
light-emitting diode

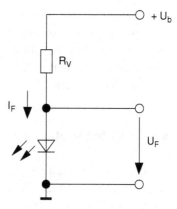

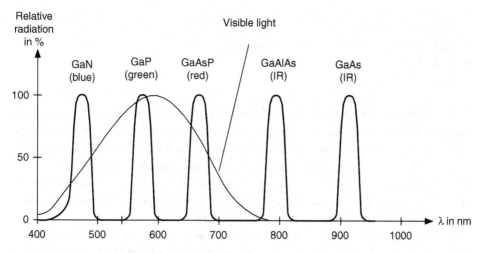

Fig. 4.27 Spectral emission of light-emitting diodes: The decisive factor is the semiconductor material used

The series resistance R_v is calculated from

$$R_v = \frac{U_b - U_F}{I_F}$$

The forward current I_F is essentially determined by the diameter of the LED. The emission efficiency of LEDs for the visible range is very small, a maximum of 10%. Figure 4.27 shows the spectral emission of the different light-emitting diodes.

The luminous intensity I of a light-emitting diode is obtained by multiplying the measured radiant intensity by the absolute eye sensitivity. The wavelength of the emitted radiation must be known very precisely. For series measurements, a calibrated silicon photoelement with a special color filter is used to simulate the red edge of the eye sensitivity curve. The BPW20 sensor cell is used as the photo element. This has a linear short-circuit current characteristic even at the lowest irradiation levels. Because of the low radiation flux of light-emitting diodes compared to IR diodes and because of the color filter, the current here is only a few milliamperes. Therefore an operational amplifier with FET input is used here.

4.3 Optocoupler

An optocoupler is used for galvanic isolation, but the logical connection of circuits. An infrared LED is used as a light transmitter, a phototransistor is usually used as a light receiver. The aims of the development are here:

- High coupling factor
- High cut-off frequency or short response time
- High insulation voltage
- Production-ready assembly

Depending on the application, there may be additional requirements, e.g., regarding linearity, transmission range, or stability. As already mentioned, the technology of the optocoupler is primarily construction and housing technology. Today they usually have hermetically sealed plastic housings. The wiring of the connections is also more or less determined by the application, with the restriction that a certain minimum distance between the external connections is necessary to achieve insulation voltages in the kilovolt range. Figure 4.28 shows the structure of an optocoupler.

A high coupling factor requires the use of IR emitters with high efficiency and photo-transistors with high infrared sensitivity. Besides, it must be ensured that the light emitted by the emitter falls as completely as possible on the phototransistor. This is done, for example, by using the light guide principle or by bundling the beams with lenticular

Fig. 4.28 Principle of an optocoupler

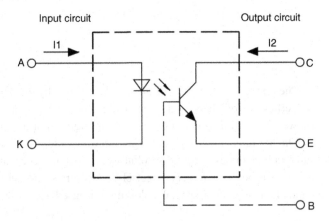

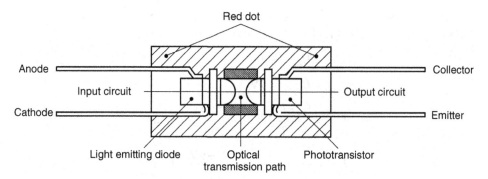

Fig. 4.29 Cross-section through an optocoupler

Fig. 4.30 Circuit for operating an optocoupler

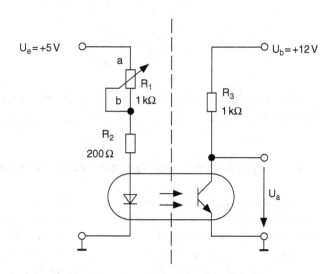

elements. In this way, the radiation can be collected almost completely, even with relatively large emitter-receiver distances, so that a high insulation voltage is guaranteed at the same time as a high coupling factor. Figure 4.29 shows the cross-section through an optocoupler in a DIL package.

Example: Within which limits does the output voltage U_a change if the switching of an optocoupler shown in Fig. 4.30 is shifted from a stop a to stop b?

Calculation of the input currents $I_{F(a)}$ and $I_{F(b)}$:

$$U_F = 1.3\,\text{V} \quad (\text{from the datasheet})$$

$$I_{F(a)} = \frac{U_e - U_F}{R_2} = \frac{5\text{V} - 1.3\text{V}}{220\Omega} = 16.8\,\text{mA}$$

$$I_{F(b)} = \frac{U_e - U_F}{R_1 + R_2} = \frac{5\text{V} - 1.3\text{V}}{1k\Omega + 220\Omega} = 3\,\text{mA}$$

Calculation of the output voltage U_a. The resistor R_3 has been defined with 1 kΩ and the following values result in the characteristic curve field:

$$U_{a(a)} = U_{CE(a)} \approx 2V$$

$$U_{a(b)} = U_{CE(b)} \approx 9V$$

$$I_{C(a)} = \frac{U_b - U_{CE(a)}}{R_3} = \frac{12V - 2V}{1k\Omega} \approx 10mA$$

$$I_{C(b)} = \frac{U_b - U_{CE(b)}}{R_3} = \frac{12V - 9V}{1k\Omega} \approx 1.5mA$$

$$V_{I(a)} \approx \frac{I_{C(a)}}{I_{F(a)}} \approx \frac{10mA}{16.8mA} \approx 0.6$$

$$V_{I(b)} \approx \frac{I_{C(b)}}{I_{F(b)}} \approx \frac{1.5mA}{3mA} \approx 0.5$$

The values $V_{I(a)}$ and $V_{I(b)}$ are within the typical value range of $V_{I(typ)} \approx 0.25$–0.7 for the simple optocoupler.

4.4   Light Barriers and Optoelectronic Scanning Systems

For decades, photoelectric switches have had a firm place in numerous industrial plants. Whenever objects have to be detected or scanned without contact, light barriers are indispensable. Most automation tasks could not be realized without them.

In recent years, the manufacturers have implemented many new ideas and significantly increased the functional reliability and ease of use of the devices. Apart from the principle, today's highly developed photoelectric switches have little in common with their ancestors. They also have a significantly higher life expectancy and are insensitive to vibration.

Today, only light-emitting diodes are used as optical transmitters instead of conventional incandescent lamps. For very high demands on beam bundling, laser diodes are also used—either in the visible or infrared range. Both can be operated in alternating light mode, making the entire system insensitive to ambient light. The receiver only registers the alternating light of its transmitter.

In practice, a distinction is made for photoelectric switches between the designs listed in Table 4.4.

Table 4.4 Overview of optoelectronic systems according to DIN 44030

Optoelectronic sensors		
Disposable systems	Reflection systems	Pushbutton systems
External radiation receiver	Reflection light barrier	Autocollimation scanner
Transmitter-receiver-separated light barrier	Reflection light grid	Angular scanners, luminescence scanners
Through-beam light grid, through-beam light curtain	Reflection light curtain	Row button

Fig. 4.31 Design of a through-beam light barrier

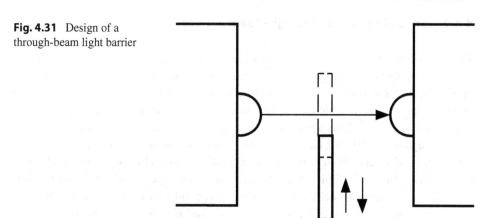

4.4.1   Through-Beam Sensors

In the case of a through-beam light barrier, the transmitter and receiver are spatially separated from each other (Fig. 4.31). The transmitter is aligned so that as much of its light as possible falls on the receiver. The receiver can clearly distinguish the received light from the ambient light. If the light beam is interrupted, the output switches on, off, or over—depending on the version.

To ensure safe operation, the transmitter has a "radiation lobe" that over-radiates the receiver. Equivalent to this, the receiver has a "receiving lobe", which is also selected to be larger. This means that even if the transmitter and receiver are not optimally aligned, the radiation flux is still sufficient. The active area is only within the light barrier between transmitter and receiver, but the radiation or receiving lobe is larger, as Fig. 4.32 illustrates.

If an optical fiber (glass or plastic fiber) is mounted in front of the transmitter and receiver, the photoelectric sensor is given an "extended eye". Since the fibers have very small dimensions and are flexible, photoelectric switches can be implemented even in places that are difficult to access. They are also free of electrical potential. They can therefore also be used, e.g., in potentially explosive areas and high-voltage systems. By selecting appropriately thin fibers, even the smallest objects can be detected.

Fig. 4.32 Optical characteristics of a through-beam light barrier

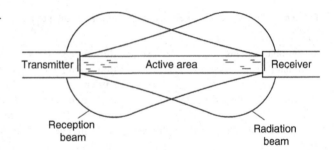

4.4.2 Special Forms of Through-Beam Sensors

Table 4.4 shows the through-beam systems, including the light barrier with the separate arrangement, the external radiation receivers, the through-beam light grids, and the through-beam light curtain. The external radiation receiver is a through-beam system without a transmitter, in which the sun or an existing lamp is the light transmitter. If an object interrupts the path between transmitter and receiver, the receiver reacts.

A through-beam light barrier works in the "one-dimensional" beam range. The stringing together of several through-beam sensors and their logical combination results in a through-beam light grid. This allows larger areas to be covered in a grid-like manner. Instead of numerous photoelectric switches, a through-beam light curtain can also be implemented with only one photoelectric switch. The light beam from the transmitter is diverted via numerous mirrors before it reaches the receiver.

4.4.3 Reflection Light Barriers

With retro-reflective sensors, the transmitter and receiver are combined into one unit and arranged on one side of the light path. A mirror is located on the opposite side (Fig. 4.33).

One measure to increase safety is the installation of polarization filters. The pulsed light from the transmitter diode is focused through a lens and directed via a polarization filter onto a reflector. Part of the reflected light reaches the receiver via another polarisation filter. The filters are selected and arranged so that only the light reflected by the reflector reaches the receiver, but not that of other objects in the beam area. This considerably increases the range. If the beam path from the transmitter via the reflector to the receiver is interrupted, the output switches.

In the case of the light barrier in Fig. 4.33, the transmitting and receiving optics are located close together. Figure 4.34 shows the optical characteristics of this light barrier.

The range for a reflex light barrier corresponds to the distance between the device and the reflector. In practice, however, a distinction must be made between operating range and limiting range. Photoelectric sensors whose technical data do not include this distinction

Fig. 4.33 Design of a reflex light barrier

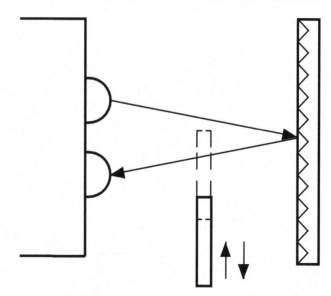

Fig. 4.34 Optical characteristics of a reflex light barrier

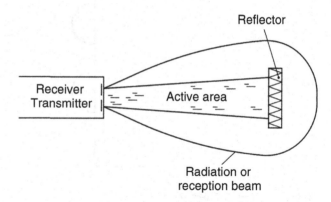

are generally not suitable for long-term use for the specified ranges, because, in practice, the soiling of lenses, the aging of some components, and also the non-optimal alignment of the system must be taken into account. For greater distances, a triple mirror is useful instead of a simple plane mirror, as this does not have to be aligned, but always reflects the light in the same direction from which it comes. It consists of three mirror surfaces at right angles to each other (cube corner). Figure 4.35 shows how it works.

In practice, such triple mirrors or triple reflectors are not composed of three individual mirrors but are manufactured as a wholly transparent body (glass or plastic), with the reflections taking place as total reflections. In many cases, such triple surfaces are also composed of several small triple elements, which are then usually injection-molded from plastic or embossed in foil ("cat's eye").

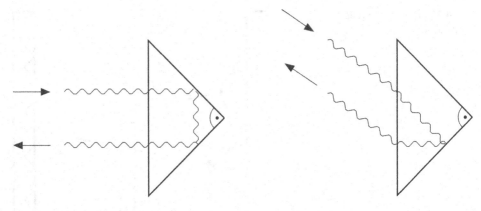

Fig. 4.35 Operation of a triple mirror

Fig. 4.36 Design of a diffuse sensor

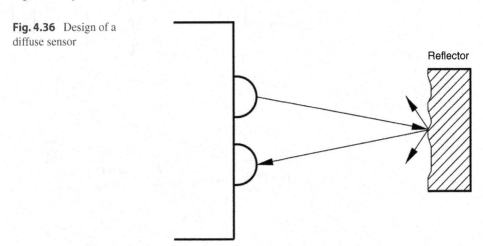

Reflector

4.4.4   Retro-reflective Sensors

Here, the pulsed light from the transmitter diode strikes an object of any shape and color. It is diffusely reflected from this object, and part of it reaches the light receiver located in the same housing. If the reception strength is sufficient, the output switches. The attainable range depends on the light intensity of the transmitter, the size, color, and surface quality of the object, and the sensitivity of the optical receiver. This can be changed within wide limits with a built-in potentiometer. It must be set so that the receiver responds reliably when an object is present but drops out again after removal.

If there are reflecting surfaces in the vicinity of the object, their disturbing effect can be reduced by using a polarization filter provided for this purpose. Figure 4.36 illustrates how a diffuse sensor works.

In principle, light barriers today are very insensitive to ambient light due to the use of alternating light. Nevertheless, there is an upper limit for the intensity of external radiation, which is called the extraneous light limit. This is measured as illuminance on the light entry surface. It is specified for sunlight (unmodulated light) and lamp light (light modulated at twice the mains frequency). At illuminance levels above the respective ambient light limit, the safe operation of the devices is no longer possible.

In an optical proximity switch, the transmitter and receiver are located in the same housing, and the object to be detected itself acts as a reflector. The active area in Fig. 4.37 is the zone in which the switch responds to an object. The size of this area depends very much on the color, surface, and size of the object being scanned. In Figs. 4.36 and 4.37 the switching curves for the respective proximity switch are drawn. The special scanning ranges for the diffuse reflection light scanners are achieved with the specified areas using matt white standard paper. For other surfaces, the correction factors listed in Table 4.5 must be observed.

The functional reserve is the excess radiant power that falls on the light entry surface and is evaluated by the light receiver. Due to soiling, changes in the reflection factor of the object, and aging of the transmitter diode, it can decrease over time, so that safe operation can no longer be guaranteed.

Fig. 4.37 Optical characteristics of a proximity switch

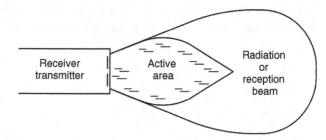

Table 4.5 Correction factors for diffuse sensors

Standard values	
Test card (reference value)	100%
White paper	80%
Grey PVC	57%
Printed newspaper	60%
Lightwood	73%
Cork	65%
White plastic	70%
Black plastic	22%
Neoprene, black	20%
Car tyres	15%
Raw aluminum sheet	200%
Black anodized sheet aluminum	150%
Aluminum matt (brushed)	120%
Steel INOX polished	230%

Some photoelectric sensors, therefore, use a second LED which lights up when 80% of the available range is used at most. Besides, there are versions where this signal is switched to one of the outputs. In this way, it is possible to detect in good time that the device is no longer sufficiently reliable.

4.4.5 Detection of Shiny Objects

High-gloss objects such as glass surfaces or reflecting metal parts can easily be overlooked by standard retro-reflective sensors. Although the objects interrupt the light path to the reflector, they often throw the transmitted light back to the receiver, so that the receiver receives light in the same way as an uninterrupted beam of light. When retro-reflective photoelectric sensors work with polarized light, as shown in Fig. 4.38, these problems are avoided.

Natural light is unpolarized and does not prefer any particular plane of vibration. If you send it through a polarization filter, it only lets the oscillation part of a certain plane of polarization pass. After that, the light only oscillates on one plane. With a second polarization filter, the intensity of this polarized light can be varied continuously. If both filters have the polarization planes in the same direction, all light can pass. However, if one

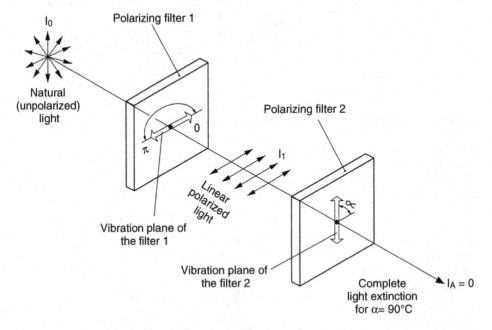

Fig. 4.38 With two polarization filters, the intensity of polarized light can be continuously varied. When the plane of polarization of the second filter is perpendicular to that of the first, no light passes through

filter is turned, the intensity of the transmitted light is reduced. At a twisting angle of 90°, no more light can pass through.

In practice, it looks like this: The photoelectric sensor manufacturer equips its transmitter and receiver of the retro-reflective sensors with one polarization filter each, which is integrated into the optics. The polarisation planes of these two filters are perpendicular to each other. If the linearly polarized light is reflected by a reflecting object, the receiver is "blind" to this, because in this case the original polarization plane of the transmitted light is retained. The reflector, on the other hand, depolarizes the light; it now has a component with a polarization plane rotated by 90°. The receiver recognizes only this light from the reflector, and the reflector cannot be deceived by the "wrong" reflected light.

4.4.6 Photoelectric Proximity Switches with Background Suppression

A concrete application for diffuse sensors is for example warehouses. Storage space is precious, which is why more and more double warehouses are being found in practice, where two pallets stand one behind the other. To reliably detect the rear pallet, the diffuse reflection light scanners must have a long-range. Recently, manufacturers have started offering devices for these distances as well, in which error signals due to background reflections are virtually eliminated. A special optical system limits the range to a defined area.

The working conditions are not always ideal for the probes. The difficulties are usually related to their mode of operation, as they evaluate the light that an object itself reflects. Often, however, the space behind the object is not free, because there may be a shiny machine part there. In this case, a normal diffuse reflection light scanner can't distinguish whether the reflected light is reflected back from the object or the background. An object lying on a conveyor belt can only be detected from above with such a device if the contrast to the belt is sufficiently high.

With diffuse reflection light scanners with background suppression, on the other hand, the view is limited to a defined distance range. Everything behind it is ignored. The border between detection and non-detection is extremely sharp and depends very little on the color and reflective properties of the surfaces. Even objects with the lowest reflectivity, e.g., black or matt materials, can be detected without difficulty.

The exact scanning range is not always fixed when the user decides on a particular probe. Therefore, it is often a great advantage if the device can be adjusted to the desired distance on-site. Some manufacturers, therefore, offer adjustable buttons with background suppression as a program supplement.

A diffuse reflection light scanner with background suppression works with two receiver elements, the near element (a) and the far element (b), as shown in Fig. 4.39. If the reflex plane approaches, the light spot moves from the far element to the near element. The signal of the near element is thus increased. A comparator compares the signals of both receiver elements and activates the output of the probe when a certain threshold value is reached.

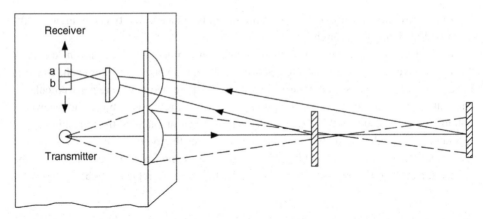

Fig. 4.39 Diffuse reflection light scanners with background suppression work with two receiver elements, the near element (**a**) and the far element (**b**)

To meet the different requirements in practice, the manufacturers offer a whole range of such probes with different ranges. The devices are therefore just as suitable for distance diagnosis in the millimeter range as for long distances in meters.

Overall, the adjustable buttons with background suppression cover a detection range from 50 to 2000 mm. Depending on the application, you can choose between maximum ranges of 300, 800, and 2000 mm. With this large scanning range, "compartment occupied" detection is also possible in double pallet warehouses.

4.4.7   Drill Breakage Control Using Light Barrier

Light barriers in designs suitable for mechanical engineering have long been standard for use in metalworking. Housed in robust metal housings, they do not resent harsh environmental conditions so quickly and reliably fulfill their functions here as well. If things do get too "rough", this is usually no problem: photoelectric switches can bridge relatively large distances and can often be easily mounted outside the danger zone. For some makes, the sensors are also available in protection class IP65 (dust-tight and splash-proof). A waterproof connector housing enables easy mounting.

Missing or broken-off tools are a considerable source of danger in metalworking. They can destroy subsequent tools or expensive workpieces and in serious cases even damage the machine tool. Automated processes can hardly be monitored with the human eye. For example, no employee can stand next to a flexible production cell and monitor the tool change, often with drills that are only millimeters thick, with "Argus eyes". Reliable tool breakage monitoring is therefore essential. Figure 4.40 explains how this can be realized with the help of a light barrier.

Fig. 4.40 Optical tool breakage detection using light barrier

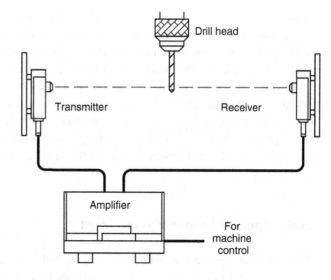

Fig. 4.41 Optical tool breakage detection using a reflex light barrier

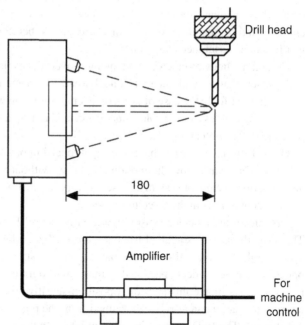

Optical monitoring is most useful for drills and taps because a light beam cannot bend, unlike a mechanical flag switch. It also works over long distances; drill tips with a diameter of 2 mm can still be easily detected.

Today's light barriers can be easily integrated into the program of a programmable logic controller (PLC). Measured against the machine times, the measuring time of the light barrier is negligible. Figure 4.41 shows an optical tool breakage detection using a reflex light barrier. This allows the scanning of drill tips even with very small diameters. The light beam emitted by the transmitter is reflected by the tool and detected by the receiver. If the tool is missing or broken, the light beam is no longer reflected. This information is transmitted to the machine control system via the amplifier.

The two light barriers can be used as "eyes" for tool breakage detection. These check with each change whether the drill is undamaged. In case of an error message, the machine change now automatically replaces the tool; you do not have to switch off the system for this. To prevent coolant and drilling chips from contaminating the light barrier optics, it is advisable to install a small slider in front of the light barrier, which is opened with each measurement.

4.4.8   Optical Distance Measurement

The faster today's production plants work and the higher the demands on production quality increase, the less the human eye can inspect the objects; it is much too slow and inaccurate. Dimensions, contours, or profiles of the objects are therefore optically recorded, checked, and compared. Automatic measurement is therefore becoming increasingly popular in quality assurance and control.

Optical methods offer several advantages over mechanical solutions. They work without contact, that is, without mechanical wear. The accuracy remains constant over the entire service life. In the case of non-contact measurement with LED or laser light, surface reflection also plays a minor role in many areas. The surface may be sticky, touch-sensitive, or easily deformable.

Optical distance sensors have great practical importance. In contrast to mechanical solutions, they can follow the motion sequences without inertia. In many cases, they are also superior to camera systems; these are very good at detecting contours, but no elevations or depressions in the direction of view.

Optoelectronic distance sensors work partly with LEDs and partly with laser diodes. They usually use the so-called triangulation method, which is explained in Fig. 4.42. A spot of light is projected onto the object to be measured with the aid of the transmitter optics; its surface reflects part of the transmitted light diffusely towards the sensor. The receiver optics image this spot on a position-sensitive detector (PSD) element. The distance of the target can also be determined from the point of incidence of the light spot on the element. The PSD element is a photodiode with two current outputs.

The ratio of the two currents to each other depends on the position of the incident light spot. The control electronics generate an analog signal proportional to the distance. Since the light position and not the light intensity is used for distance measurement, the method is largely independent of the reflection properties of the object surface. Even objects with low reflectivity (at least 10%) can be measured reliably. An internal intensity regulator automatically adjusts the power of the emitting source to the respective reflectivity.

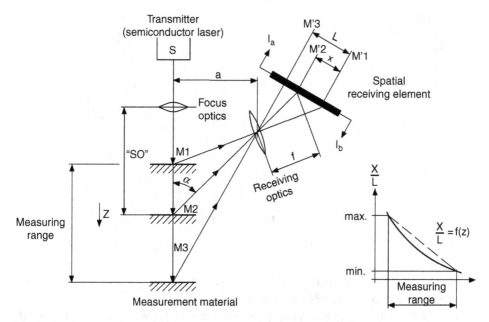

Fig. 4.42 Non-contact and optical distance measurement using the triangulation method

Distance sensors with a laser diode as a transmission source can be used for all applications where high accuracy and speed are required. One manufacturer, for example, has developed a type that covers the measuring ranges from 2 to 100 mm with four different optics. The resolution is 0.05% of the measuring range, the maximum measuring frequency is 10 kHz. With a temperature coefficient of 0.01%/K, the temperature drift of the measured value should normally be less than the thermal expansion of the target. However, in the interests of accuracy and reproducibility, the operating temperature must be kept constant within the permissible range between 5 and 40 °C. This makes the measuring system ideally suited for a wide range of different applications in industry and research for thickness or layer measurements and profile scanning of plastics, sheet metal, wood, or textiles.

Another important field of application for these precise measuring systems is the automotive industry. The laser distance sensor is suitable for unbalance measurements as well as for vibration measurements or machine monitoring. Other typical applications include a dynamic track play measurements on railways, the stroke detection of dynamic, mechanical processes, or counting objects with variable transport heights.

A typical laser distance measurement system consists only of the sensor in cigarette pack size and a power supply that can be installed up to 10 m away from the sensor. A built-in LED bar graph in the power supply provides information about the current position of the target and serves as an alignment aid. At the output, the sensor provides a distance proportional current signal from 4…20 mA. Measuring adapters for output voltage

Fig. 4.43 Displacement
measurement using the
triangulation method

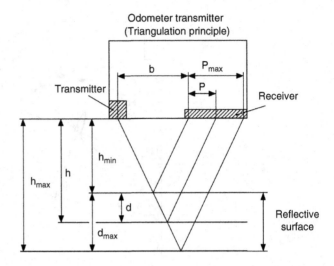

adjustment with low-pass filters and offset function for shifting the measuring range and
for zero adjustments are available as options.

If the high resolution, accuracy, and speed of the laser distance sensors are not required,
there is a low-cost alternative with the LED transmitter element, which also works according to the triangulation method. However, the measurement frequency of the device is only
at 100 Hz.

With a highly reflective surface—according to Fig. 4.43—the measuring principle is
also possible. This variant of triangulation is based on the fact that in an isosceles triangle
with a given base b and the angle f between base and leg, the triangle height is also unam-
biguously determined. Such an isosceles triangle is created by directing a light beam of the
LED against the reflecting surface (not perpendicular to it), reflecting it from there according to the law of "angle of incidence equals the angle of reflection" and then falling on the
receiver. The transmitter-receiver connection forms the triangle base b, the light beam
before and after reflection one leg each. The triangle height h (starting from the base) now
indicates the desired distance.

If you change the distance to the reflecting surface, the base and the height in the triangle, that also changes the distance. To determine the height from the respective base, one
considers that with varying distance, similar triangles are formed, which all meet the
condition

$$\frac{h}{b} = c = \text{constant}$$

where the constant ratio c between height and base depends only on the angle of radiation
φ and the value

$$c = \frac{\tan \varphi}{b}$$

accepts. However, the trigonometric relationship no longer appears in the practical calculations, since the proportionality between b and h is sufficient for the relationship,

$$d = c \cdot p$$

i.e., the measuring range (0 to p_{max}) of the receiver can be transferred linearly to the distance measuring range (0 to d_{max} or h_{min} to h_{max}). Thus the resolution of the receiver is also directly transferred to the resolution of the distance measurement itself. The resolution itself is specified with <1‰.

For range-variable measurements, the angle of radiation φ can be selected appropriately and the constant c can be calculated either by reference measurements or if the angle φ is sufficiently accurate. If, on the other hand, it is known in advance in which range the distances are to be measured, it is advantageous to combine transmitter and receiver into one unit. Today's distance sensors are accommodated in a compact housing and are suitable for measurements at close range with an angle of radiation φ of 70° (50...70 mm). The applications are by no means limited to pure distance measurement, but also allow recording the movement of an object, especially vibrations. Figure 4.44 shows a vibration measurement with an optical distance sensor.

The internal structure of an optical distance or displacement sensor is shown in Fig. 4.44. An infrared semiconductor laser with a light output of 1.5...2 mW is used as the light transmitter, and a PSD element is used as the position-sensitive receiver, which detects the position of impact of the light spot with very high accuracy. The generated analog signal is first processed in an evaluation unit and digitized with an AD converter.

In further processing, the actual distance is calculated from the measured values. Depending on the concrete task at hand, the values can then be forwarded to other system components. A dynamic recording of the distance data with a pre-settable sampling frequency (cut-off frequency 50 kHz) makes it possible to record the movement over a certain time interval. If the movements show a certain periodic behavior, a subsequent

Fig. 4.44 Vibration measurement using an optical distance sensor

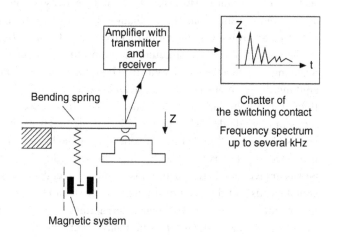

graphical representation as well as FFT-analysis can be used to provide information about the frequencies involved.

4.5   Optical Angle of Rotation and Position Detection

In the automated machine sector—especially with robots—geometric variables such as absolute position, distance traveled, angle of rotation, or speed of machine parts or workpieces must be continuously recorded. These tasks are performed by corresponding measuring systems that are connected to the machine axes or other moving parts in a suitable manner.

The sensors used here very often operate by optoelectronic means. The measured values obtained are fed to the computer, which compares the actual position values with the setpoint values specified by the program and initiates appropriate control measures for motors and other actuators.

The rotary encoders in particular have become very widespread. These can be distinguished between absolute and incremental versions.

4.5.1   Absolute Rotary Encoder

A so-called absolute rotary encoder is used to detect the angular position of a rotating part. In principle, it is a type of AD converter that converts a mechanical analog value into a digital electrical signal—a word consisting of several bits that is written in parallel at the encoder output and represent the absolute angular value. The advantage of this type is that the value is immediately available when the power supply is switched on—in contrast to incremental rotary encoders, which will be discussed later and which must first approach a reference mark. This makes the measuring system largely insensitive to disturbances such as power failures. However, this is at the expense of higher costs for measured value acquisition, transmission, and evaluation. For this reason, absolute encoders are used in industrial and robotic applications that require particularly high safety.

Figure 4.45 shows the structure of an absolute encoder with downstream electronics. Scanning is contactless and wear-free by optical means. The light of an infrared light-emitting diode shines through a rotatable code disk made of glass and a fixed aperture. Behind it, a light-dark pattern is created, which is converted into electrical signals by a series of photodiodes. The code is only repeated after a complete rotation. Such designs are also known as "single turn" encoders. How the binary words change from step to step is defined as "code type". The "normal" binary code based on the principle of dual numbers is unfavorable here because it is a multi-step code in which several bits often change simultaneously during the transition from one step to the next. Due to possible mechanical manufacturing tolerances between different tracks and electrical propagation delays in the individual channels, slight time shifts may occur, which then lead to incorrect digital data

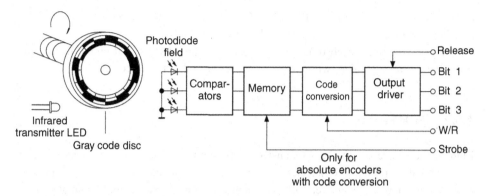

Fig. 4.45 Design of an optical absolute encoder with downstream electronics

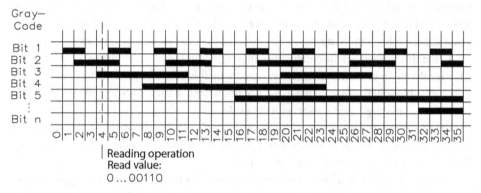

Fig. 4.46 Structure of a one-step code according to Gray

for a short moment (so-called "glitches"). Much safer here is a "one-step" code in which only one bit changes at a time during the transition from one binary word to the next—even at the zero-crossing. If the transition on the code disc is not exactly in the correct angular position, at most the switching point shifts slightly from one binary word to the next. However, the correct value is always read, and no incorrect intermediate combinations occur. This condition of one-step operation is fulfilled by the so-called Gray code of Fig. 4.46. It is used with most absolute encoders. The intermediate storage and a subsequent electrical code conversion ensure that edge delays do not result in false information. However, this requires a relatively high electronic effort.

By inverting the most significant bit the gray code can be reversed very easily. This means that you can freely decide during processing whether the code increases or decreases in numerical value when it is turned clockwise. This possibility is achieved by the reflectivity of the one-step code, which is also valid for the symmetrically capped Gray code. This is a certain section of a complete Gray code; it allows us to apply any even-numbered step divisions without losing the one-step capability and the reflectivity. After the code conversion from the symmetrically capped Gray code to the natural binary code, the latter is afflicted with an offset, which can be eliminated by software.

The principle used in absolute rotary encoders is also used to determine linear positions. In this case, a "ruler" with Gray-coded sections takes the place of a disk.

4.5.2   Fork Sensors

Light barriers have already been extensively reported in Sect. 4.4. A miniaturized version of the so-called forked photoelectric switch is used for speed and incremental angle of rotation detection: The so-called forked photoelectric switch is a through-beam photoelectric switch in which the optical transmitter and receiver are only a few mm apart and are located in a common housing. Additional lenses and apertures optimize the beam path and thus improve the resolution. A disk with translucent and opaque segments runs through the slot (Fig. 4.47). Figure 4.48 shows one variant: Here, the transmitter and receiver do not point at each other, but in the same direction; the disk has reflecting and black segments.

As the disc rotates, the luminous flux impinging on the receiver changes periodically. This results in an electrical alternating signal which is amplified and formed into a digitally evaluable square-wave signal using a Schmitt trigger. Figure 4.48 shows the principle of the forked photoelectric sensor with the evaluation of electronics.

The frequency of the signal is proportional to the speed. By counting the pulses per time unit, a speed sensor is obtained (Fig. 4.49). Its measuring range is in principle only limited by the reaction speed of the photodiode.

Such an arrangement with only one light barrier does not yet provide information about the direction of rotation. A change of direction is not detected and leads directly to

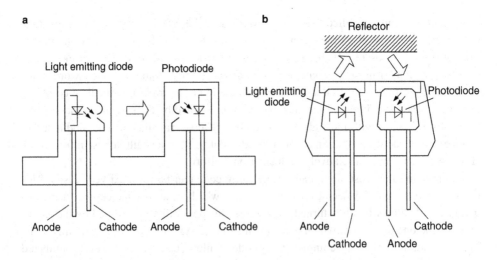

Fig. 4.47   Principle structure of transmissive (**a**) and reflective (**b**) fork sensors

Fig. 4.48 Speed measurement
with a forked light barrier

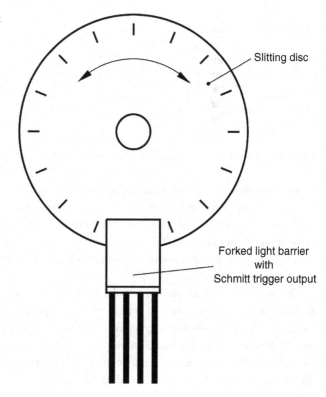

Slitting disc

Forked light barrier
with
Schmitt trigger output

Fig. 4.49 Structure of a
two-channel encoder with a
pulse diagram for detection of
the direction of rotation

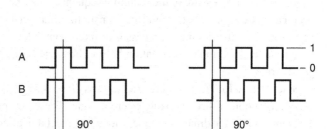

incorrect results. The two-channel light scanning as shown in Fig. 4.51 with two receivers
E1 and E2, a raster R, and a transmitter S with as parallel a luminous flux P as possible in
the direction of the receivers is a remedy here. Certain geometric relations must be
observed for proper function.

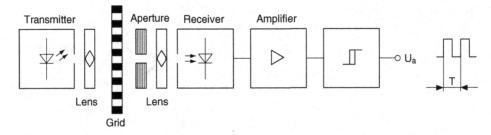

Fig. 4.50 Principle of the forked light barrier with the downstream evaluation electronics

The direction of rotation results from the phase shift between the two output signals. The ratio of the segment distance on the disc to the distance between the photodiodes is made so that the phase shift between the two signals is just 90°. Figure 4.50 shows such a two-channel rotary encoder. If channel B already has a "1" signal on a rising edge in channel A, the axis rotates clockwise; if channel A has a "0" signal on a rising edge, the axis rotates counterclockwise.

The width of each of the two similar photosensitive elements E1 and E2 is marked B. The distance between the two is marked A; the opaque segment of the scanner has the width U and the transparent segment has the width D. Reliable direction detection with parallel light output is now based on the phase shift φ of the two signals Q_A and Q_B, which are individually processed according to the forked photoelectric sensor principle (below in Fig. 4.51). This only has to correspond to the relation

$$0 < \varphi < \pi$$

are sufficient, which is guaranteed by the following geometry condition:

$$A + B < U, D$$

The further processing of the scanning sequences obtained in this way has so far mostly been carried out in a costly and time-consuming manner using logic components or the peripheral circuits of a microprocessor or microcontroller. The problem here is always to have reliable directional information before the respective counting pulse and to exclude pulse errors.

Since 1991, forked photoelectric sensors (Fig. 4.51) have been available which can solve this problem directly using internal logic. These components contain an integrated evaluation logic for direction recognition. The mechanical design is identical to the conventional photoelectric sensor, but the internal logic has been considerably improved. The transmitter is a GaAlAs LED, whose luminous flux is directed quasi-parallel to the receiver using a sharpened lens.

Errors due to age-related different changes in the light output, as can occur with fork light barriers with two transmitters, are thus excluded. The receivers with daylight blocking filter, which can be operated over a wide range of output levels, make a readjustment

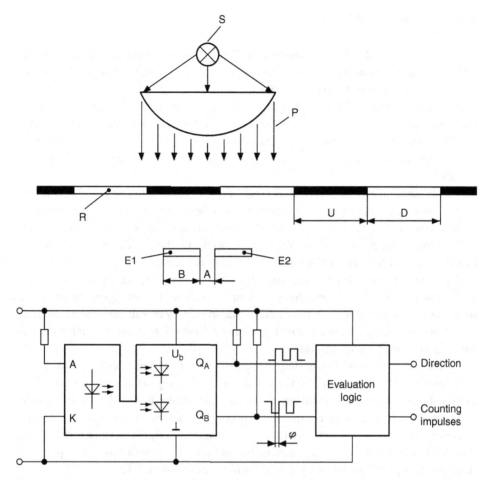

Fig. 4.51 Structure of a two-channel forked photoelectric sensor: the functional structure at the *top*, the electrical part at the *bottom*

of the transmitters unnecessary. The receiver assembly consists of two narrow, closely spaced photodiodes that allow any grid spacing down to 0.85 mm (1/30 inch) for a certain tolerance range.

In addition to the photodiodes, these fork sensors contain an integrated receiver circuit with all necessary signal conditioning components. The logic for generating the signals for direction and increment is designed in such a way that the direction information is always reliably output before the counting pulse (for example 1 µs). The width of the counting pulses is independent of the light-dark alternating frequency and is typically 10 µs. Counting errors caused by light barriers, e.g., during changes of direction or oscillating movements, are excluded as long as they do not fall below the grid spacing of 0.85 mm and the recommended geometrical arrangement is also observed.

4.5.3   Incremental Encoders

So-called incremental encoders consist of a forked light barrier and a disk with translucent and opaque or reflecting and non-reflecting segments. First of all, they have only one data output at which trapezoidal to approximately sinusoidal electrical pulses appear when the encoder axis is rotated. With an electronic counter, it is now possible, to sum up, these pulses and thus determine the angle of rotation passed through. To also record the direction of rotation, another output is required, which supplies pulses phase-shifted by 90°. A simple logic circuit or a program then recognizes the direction of rotation from these two signals.

The disadvantage of the incremental rotation angle sensor is that after a power failure or when switching on, there is no information about the current position. A reference mark is still required for this. For this purpose, they have an additional scanning track that only provides one pulse per revolution. When setting up the machine, the reference point must be traversed once to start counting from a defined initial value.

The simplest pulse shaper electronics consists of three comparators with which the initially analog signals are converted into binary signals for further digital processing. If the signals of the photoreceivers have a clean sine wave, they contain much more information. Thus, a period can be interpolated, that can be divided into a certain number of measuring steps to increase the resolution.

Correct mounting is of decisive importance for the accuracy and interference immunity of the encoder. This applies equally to the mechanical coupling to the machine axis and the electrical interface to the subsequent electronics and further to the PLC or PC system. Flexible connecting elements are inserted between the machine and encoder axes to compensate for axial misalignment, angular error, and radial misalignment when installing the encoder. For normal requirements, a metal bellows coupling is sufficient, while precision diaphragm couplings are already required for high-resolution encoders.

For interference-free electrical signal transmission, the pulse outputs are equipped with differential line drivers. The signal for the data transmission is transmitted directly and negated, making it largely insensitive to common-mode interference interfering with the cable. An RC combination ensures the correct terminating impedance to prevent reflections on the cable. For wire break monitoring, special resistors are arranged at the input of the operational amplifier to prevent the receiver from switching through in the event of a fault.

The principle is also used to detect translational movements. A grid ruler with translucent and opaque strips takes the place of the disc. The signal evaluation is the same. Reference marks are also required here, which must first be approached after switching on.

4.5.4   Signal Evaluation

The most important static characteristic value is the measuring step, which is determined by the number of lines on the graduated disk on the one hand and by the electronic signal evaluation on the other. The line count of the encoder can be in a wide range from 50 to 36,000, depending on the application. The electronic evaluation essentially comprises two different procedures: Edge evaluation and interpolation.

Edge evaluation takes advantage of the fact that one period comprises exactly four edges (two of the non-shifted and two of the 90°-shifted signal), which are primarily used for direction detection but are also suitable for refinement by an edge evaluation circuit by a factor of 2 or 4. The signal evaluation is carried out directly, that is, without interpolation, at "coarse" resolution down to 0.1° (corresponding to 3600 steps per revolution). A further increase in resolution is possible by interpolation of the sinusoidal scanning signals, subsequent square-wave conversion, and edge evaluation. Usually, the division of a period into 5, 10, or 25 equal sections is chosen.

An encoder with 9000 lines, fivefold interpolation, and quadruple edge evaluation therefore provides $9000 \cdot 4 = 36,000$ steps/revolution, which corresponds to a resolution of 0.01°.

The dynamic characteristics of incremental encoders depend mainly on the properties of the photoelectric sensors and the permissible input frequency of the evaluation circuit. In the simplest case, the evaluation circuit consists of comparators. The standard value for the maximum scanning frequency is $f_{max} = 750$ kHz, for expensive special scanners it can be up to 10 MHz.

The most important parameter for the user is the maximum speed n_{max}. This can be calculated with the relationship

$$n_{max} \left[\text{min}^{-1} \right] = \frac{f_{max}}{\text{rpm}} \cdot 60$$

from the line count, where f_{max} is between 125 kHz and 10 MHz depending on the type.

The maximum speed for an encoder with 2500 lines and the maximum scanning frequency of 1 MHz is calculated as follows

$$n_{max} = \frac{1 \cdot 10^6}{2500} \cdot 60 \left[\text{min}^{-1} \right] = 24,000 \text{min}^{-1}$$

The accuracy of the encoders is determined not only by the cleanly designed mechanics but also by the angular scale, that is, the precision of the radial grid distribution. Today's graduations are manufactured using the same photolithographic process as modern semiconductor components and are therefore extremely accurate.

The remaining error is indicated as a directional deviation in angular degrees. In addition to the mechanical errors, it includes the errors of the pitch, the subdivision errors, and errors of the coupling used in the measurement. For rotary encoders with less than 9000 lines, the direction error is less than the resolution with fivefold interpolation and quadruple evaluation: This error is thus within a range of ±1/20 of the grating period and can be expressed by the simple relationship

$$\text{Maximum directional deviation} = \pm \frac{18}{\text{rpm}}$$

in angular degrees. For higher line counts, this deviation is given as an absolute measure to be compatible with the system accuracy.

Humidity Sensors

5

Summary

A very important measurand in practice is the humidity of gases, especially of air. Too much or too little moisture can damage not only plants but also people and disturb the function of technical equipment. Humidity sensors and the associated measuring systems help to increase well-being and safety. Today, there are numerous versions available that reliably, quickly, and economically measure the moisture in a variety of gases and also the water content in non-aqueous liquids (oil, petrol, etc.). The following specifications are in use in moisture measurement technology:

- Relative humidity in percent (% r. h.)
- Absolute humidity (grams of water per cubic meter of air, g/m^3)
- Dew point (°C)
- ppmv (parts per million by volume)
- ppmw (parts per million by weight)

While the specifications % r. h. and g/m^3 are in practice only used for gases, dew point specifications in °C and relative proportions in ppmv and ppmw are also common for liquids.

5.1   Physical Measurement Methods

There is no generally valid definition of the term moisture. The physical measuring principle used for moisture analysis often influences the definition. The following section describes moisture as it is obtained in connection with thermal (dry) determination methods.

© Springer Fachmedien Wiesbaden GmbH, part of Springer Nature 2022   309
H. Bernstein, *Measuring Electronics and Sensors*,
https://doi.org/10.1007/978-3-658-35067-3_5

The moisture of material comprises all those substances that volatilize when heated and lead to a loss of weight of the sample. The weight loss is recorded with a balance and interpreted as moisture content. Thus, no distinction is made between water and other volatile components, as is the case with a solvent, for example.

When determining moisture content, it should be noted that water can be bound differently in solids (Fig. 5.1): With increasing bond strength as

- Free water on the surface of the sample substance
- Water in large pores, cavities, or capillaries of the sample substance
- Adhesive water adhering to the surface of polar macromolecules
- The water of crystallization enclosed in lattice ions or coordinated to ions

Fig. 5.1 Type of moisture-binding

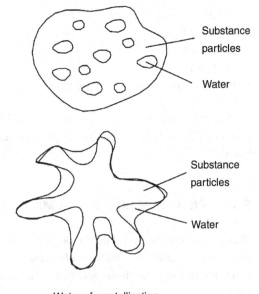

Water on the surface

Substance particles

Water

Water in pores or capillaries

Substance particles

Water

Substance particles

Water

Water of crystallization

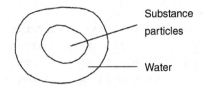

Plaster: $C_a = SO_4 \cdot 2H_2O$

The water of crystallization as well as adhesively bound water can occasionally only be separated by thermal processes with simultaneous product separation.

5.1.1   Appropriate Measurement Methods

The choice of a suitable measuring method depends mainly on the following variables:

- Requirements on the accuracy, measuring range, repeatability, sensitivity
- Type of binding of the water
- Requested information: water or moisture content
- Measuring speed
- Sample quantity
- Physical properties of the sample (e.g. Decomposition temperature)
- Costs
- Simplicity (operation or functionality)
- Legal requirements (reference methods)
- Automation capability
- Calibratability

Moisture is present in most natural products. The water content itself is rarely of interest. Rather, it indicates whether a product has certain properties that are decisive for trade and production, such as

- Shelf life
- Lump formation with powders
- Microbiological stability
- Flow properties, viscosity
- Dry matter content
- Concentration or purity
- Commercial quality (compliance with quality agreements)
- Nutritional value of the product
- Legal conformity (food regulation)

Trade and industry are interested in the dry matter content of commercial goods. The water present in the product is taken into account when setting the price. Legal regulations and product declarations define the difference between natural moisture and moisture added to the product.

It must be possible to carry out moisture measurements quickly and reliably to intervene quickly in the production process and to avoid long production downtimes. For this reason, many manufacturers today determine the moisture content of raw materials, intermediate products, and finished products directly at the production line, in line with quality assurance requirements.

5.1.2   Methods of Moisture Content Determination

The moisture content influences the physical properties of a substance such as weight, density, viscosity, refractive index, electrical conductivity, and many more. Over time, various methods have been developed to measure these physical quantities and express them as moisture content.

The measuring methods can be logically classified according to the following procedures

- Thermogravimetric
- Chemical
- Spectroscopic
- Other

The thermogravimetric methods are in principle weighing-drying methods, in which the samples are dried until a mass constancy is reached. The change in mass is interpreted as moisture released.

Drying ends when a state of equilibrium is reached, that is, when the vapor pressure of the moist substance is equal to the vapor pressure of the environment. The lower the vapor pressure of the environment, the lower the residual moisture remaining in the material in the case of equilibrium. By reducing the pressure, the ambient vapor pressure can be lowered, thus accelerating the drying conditions.

For reproducible thermogravimetric moisture determinations, the drying temperature and the drying time are of great importance. These influence the measurement result. The influence of air pressure and humidity is of secondary importance. However, these must be taken into account for high-precision analyses.

Thermogravimetric methods are classical methods for moisture analysis. For historical reasons, they are often part of legislation (food regulations, etc.).

In thermogravimetric processes, moisture is always separated following the definition given in the introduction. Thus, no distinction is made between water and other non-volatile product components.

Thermogravimetric methods are suitable for practically all thermal substances with a moisture content $>0.1\%$.

In chemical methods, there are two methods for determining moisture:

- Karl Fischer Titration
- Calcium carbide process

The Karl Fischer method is in principle used as a reference method for many substances. It is a chemical-analytical process based on the oxidation of sulfur dioxide in methanolic-basic solution by iodine. In principle the following chemical reaction takes place:

$$H_2O + I_2 + SO_2 + CH_3OH + 3RN \rightarrow [RNH]SO_4CH_3 + 2[RNH]I$$

The titration can be carried out volumetrically or volumetrically. In the volumetric procedure, a Karl Fischer solution containing iodine is added until the first trace of excess iodine is present. The amount of iodine converted is determined from the volume of the iodine-containing Karl Fischer solution using a burette. In the coulometric method, the iodine involved in the reaction is produced directly in the titration cell by electrochemical oxidation of iodide until a trace of non-reacting iodine is also present. The amount of iodine produced can be calculated from the amount of current required for this using Faraday's law.

Karl Fischer titration is a water-specific moisture determination method that is suitable for samples with high moisture content (volumetry) but also samples with water contents in the ppm range (coulometry). Originally it was designed for non-aqueous liquids, but it is also suitable for solids, provided that they are soluble or if water can be removed from them by heating in a gas stream or by extraction.

The advantages are an accurate reference method. Coulometry is also suitable for trace analysis and water detection.

As a limitation, the working technique must be adapted to the respective sample.

In the calcium carbide process, a sample of the wet substance is carefully mixed with excess calcium carbide, and the following reaction takes place:

$$CaC_2 + 2H_2O \rightarrow Ca(OH)_2 + C_2H_2$$

The quantity of acetylene produced is determined either by measuring its volume or by the increase in pressure in a closed vessel.

The process is cost-effective.

The disadvantage of this method is that calibration is necessary because not all the water contained in the sample is involved in the reaction. The formation of explosive substances hydrogen or acetylene is the reason why this method of water determination is not very common.

5.1.3   Indirect Measurement Methods

Spectroscopic moisture analysis methods are indirect measurement methods. All these methods require calibration, which establishes the relationship between the display of the spectrometer (primary measured variable) and the value determined using reference methods.

Fig. 5.2 Infrared spectroscopy
to determine surface moisture

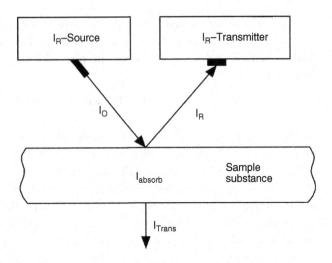

Even the quantity measured by a spectrometer is never a function of humidity alone but depends on other parameters such as density, temperature, and material properties.

The limitation is to achieve a clear correlation between humidity and physical measurand, and all other parameters must be kept constant. However, since the measured goods are never completely homogeneous and constant in their properties, all characteristic curves show a more or less pronounced scatter. Therefore, calibration requires a large number of samples that are representative of the intended application.

Infrared spectroscopy is used to determine the surface moisture. With this moisture determination method, only the surface moisture is measured. A sample is irradiated with light (electromagnetic radiation). The intensity of the reflected spectrum provides the basis for determining the moisture content. The near-infrared range used in the electromagnetic spectrum comprises wavelengths between 800 and 2500 nm. In this range, triatomic water (H_2O) has two distinct absorption bands at wavelengths 1.475 and 1.94 μm.

If a moist sample is irradiated with light of these wavelengths, part of the light is absorbed, a second part is diffusely reflected and a third part penetrates the sample (transmission). The diffusely reflected intensity (spectrum) is measured, which is proportional (non-linear) to the water concentration on the surface. Occasionally the spectrum of transmission is also evaluated for thin samples. Figure 5.2 shows the principle of infrared spectroscopy for determining surface moisture.

$$I_O = \text{Irradiated intensity}$$
$$I_R = \text{Reflected intensity}$$
$$I_{absorb} = \text{Intensity absorbed in the sample } I_R$$
$$I_{Trans} = \text{Intensity transmitted through the sample } I_R$$

Advantages are the short measuring time in the range of seconds as well as the possibility to perform multi-component analyses and real-time measurements.

The restriction is that a previous substance-specific calibration is necessary.

Microwave spectroscopy is used to determine the total moisture. Due to the extraordinarily high dielectric constant of water (DK = 81), microwaves are absorbed, reflected, and scattered in moist materials. From this, measuring moisture methods can be derived, which can be realized as transmission, reflection, or resonator methods in adaptation to the measuring task.

The transmission and the reflection method will not be described further here, as it is a modification of the NIR moisture determination method and offers similar advantages and disadvantages.

5.1.4   Laboratory Measurement Procedures of Higher Accuracy

The resonator method is a discontinuous laboratory measuring method of higher accuracy. In this method, the material to be measured is pushed into a cavity resonator and made to "oscillate" using microwaves. The existing polar water molecules absorb a part of the energy and change the microwave field. A shift in the resonance frequency and a change in the amplitude of the oscillation can be measured. This shift is not linearly proportional to the water content of the substance, but the temperature and sample weight must be known for compensation.

For both the transmission measurements and the resonator method, a substance-specific calibration of the measuring system is required. Continuous measurements can be performed with the transmission or reflection method.

NMR (Nuclear Magnetic Resonance) spectroscopy works with a high degree of accuracy. The 1H-NMR spectroscopy determines the number of hydrogen nuclei in a substance. This is then used to determine the amount of water in the sample. Two different types of 1H-NMR spectroscopy are used for moisture content determination.

In the first type, the substance under investigation is exposed to a high-frequency alternating magnetic field. The resonance behavior of the hydrogen nuclei (spin of the protons) is a measure of the water content in the substance. In the second type, hydrogen nuclei (protons) are deflected by a magnetic pulse. The rebound of the spin of the protons induces a voltage in a receiving coil. Mathematical processing of the measurement signal produces an NMR spectrum which provides information about the presence of hydrogen atoms in the sample.

The procedure has some limitations. It should be noted that NMR spectroscopy detects all H-atoms present in the sample. The hydrogen atoms that do not belong to the water molecule are also shown, but the binding conditions are visible in the spectrum. NMR signals must therefore be calibrated to the components to be measured and adjusted to the structure of the substances. However, a substance-specific calibration as in NIR spectroscopy is not necessary.

An advantage of nuclear magnetic resonance spectroscopy is its high accuracy. It detects all forms of water, regardless of the strength of the bond.

The main disadvantages are the considerable amount of equipment required and the high costs involved. The small sample size can also be disadvantageous for quality assurance applications in the industry.

With the water determination methods described in this chapter, mostly physical product properties are tested, which, with a simple composition of the sample substance, depend exclusively on the water content. These methods often require calibration and can only be used if a simple analysis matrix (the component-based composition of the sample) is available and the components present in addition to the water exclude any disturbance of the measurand.

Another method is the acquisition of the primarily measured quantity in conductometry using the electrical resistance. This is the greater, the fewer charge carriers (depending on the water content) are present for charge transport. The electrical resistance is therefore a measure of the water content of the sample.

This process is suitable for substances that have a very low conductivity in the dry state. They show a significant temperature dependence, which can be corrected by temperature measurement. There are discontinuous measuring cells into which the material to be measured must be introduced, the plug-in electrodes for measuring the conductivity must be attached and continuous measuring equipment must be available for the measurement.

Refractometry is an optical measuring method. The refractive index is measured, which (e.g. for dissolved sugar in water) is in a non-linear relation to the sugar concentration. A direct determination of water content is therefore not carried out. The measured value only defines the analysis matrix (in this case water) that is present together with the sugar. Figure 5.3 shows the mode of operation of refractometry.

This method of moisture determination is particularly important for products containing sugar. Refractometric methods can also be used for the water determination of other pure substances, such as glycol solutions.

Another measuring method is the determination via density. This method of analysis is usually only used for pure solutions, whereby the density is a measure of the concentration

Fig. 5.3 Mode of operation of refractometry

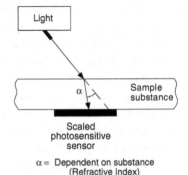

of a substance present in an aqueous solution (e.g. NaOH, sugar solutions, alcohol-water mixture). Here, too, the residual matrix of the examined substance is interpreted as water content.

The advantage of this simple method is that the results for two-substance mixtures can be read directly from tables.

This method is usually fast and the analysis can often be performed directly in the sample material. It is therefore particularly suitable for quick random sample checks and trend analyses. Its informative value about the water content is essentially dependent on the difference in density and the number of substance components. The most common methods for density determination are:

- Determination by hydrometer
- Determination using a pycnometer
- Measurement of the Archimedean buoyancy force using a balance (force gauge)
- Principle of the oscillating tuning fork

Brief description of the functional principle (Fig. 5.4) of the tuning fork: The hollow legs of a tuning fork-shaped vibrating body are filled with the sample substance. The natural frequency of the tuning fork depends on the density of the substance.

In gas chromatography, the sample mixture must first be vaporizable without decomposition so that it can then be transported in gaseous form through a separation column utilizing an inert carrier gas. In this column, the individual sample components separate due to their different boiling temperatures and intermolecular interactions between the liquid, stationary phase in the separation column, and the sample components in the mobile gas phase. The detection of the individual gas fractions emerging from the separation column is usually done by thermal conductivity (WLD).

Gas chromatography is suitable for liquid samples with a moderate solid content and a water content above 5% as well as for samples from which the water is removed by extraction.

Fig. 5.4 Principle of the oscillating tuning fork for density determination

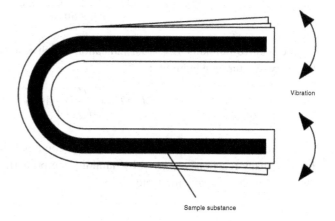

Vibration

Sample substance

The advantage of this analysis method is that several liquid sample components can be analyzed simultaneously.

The restrictions are that they can only be used for water content determinations. Also, because of their high expenditure on equipment, they can only be used if other sample components are of interest or if substances contained in the Karl Fischer titration lead to disturbing side reactions.

The method using osmometry is the measurement of the vapor pressure, which provides information about the moisture content of the sample.

5.2   Physical Relationships

To explain the different measured variables, the physical relationships will be dealt with in the following sections. In practical humidity measurement technology, a distinction is made between

- absolute humidity
- the saturation humidity
- the relative humidity

The absolute humidity H_{abs} indicates the amount of water contained in a given volume of air. It applies:

$$H_{abs} = \frac{\text{Mass of Water}}{\text{Air volume}} \left[\frac{g}{m^3} \right]$$

The saturation humidity H_{sat} indicates the maximum possible amount of water that can be contained in a given air volume. H_{sat} depends on the temperature and rises sharply with it:

$$H_{sat}(T) = \frac{\text{Maximum mass of water}}{\text{Air volume}} \left[\frac{g}{m^3} \right]$$

The relative humidity H_{rel} is an indication which results from the ratio of absolute humidity to saturation humidity:

$$H_{rel}(T) = \frac{H_{abs}}{H_{sat}(T)} \cdot 100\%$$

The indication of relative humidity is very common. The measurement is justified because many reactions triggered by humidity are primarily linked to relative humidity (rust, mold, physical condition, etc.).

5.2.1   Definition of the Water Vapor Partial Pressure

The pressure that the gas of N particles enclosed in volume V exerts on the wall can be described by the following equation:

$$dp = [b]\frac{N}{V} \cdot k_B T \quad p \;\; = \text{Pressure}$$

$$N \;\; = \text{Number of gas particles}$$
$$V \;\; = \text{Volume of the gase}$$
$$k_B = \text{Boltzmann constant}$$
$$T \;\; = \text{Temperature of the gas}$$

The atmospheric pressure of the ambient air is made up of the individual pressures of the individual components of the room air, also known as partial pressures. The air consists mainly of the following gases:

- Nitrogen (p_1)
- Oxygen (p_2)
- Noble gases (p_3)
- Trace gases (p_4)
- Water vapor (p_W)

Thus the total pressure of the atmospheric air results from the addition of the individual partial pressures:

$$p_G = p_1 + p_2 + p_3 + p_4 + p_W$$

If one now summarizes p_1 to p_4 as the total partial pressure p_L of dry air, the total pressure p_G is added to the atmosphere:

$$p_G = p_L + p_W$$

Under normal conditions $p_G = 1013$ mbar; this is the average atmospheric pressure Assuming that the water vapor contained in the ambient air exerts a partial pressure of 13 mbar, the following quantity ratios are calculated with 1 m³ air:

$$N = 2.665 \cdot 10^{25} \;\; \text{Molecules of dry air}$$
$$n = 3.461 \cdot 10^{23} \;\; \text{Water molecules}$$

If you normalize this to 10^6 molecules of the total air, you get about 12,835 molecules of water vapor, that is, the air has a moisture content of 12,835 ppmv (parts per million by volume). If a gas volume is now compressed, i.e., compressed, the individual partial pressures inevitably increase. Mathematically, the moisture content in ppmv in gases then results from the ratio of the water vapor partial pressure p_W to the total pressure of the system p_T multiplied by 10^6:

$$\text{ppmv} = \frac{p_W}{p_T} \cdot 10^6$$

5.2.2 Dew Point

Another very important parameter in humidity measurement technology is the dew point temperature. A gas cannot absorb an unlimited amount of water vapor at a certain temperature. As soon as the so-called saturation value is reached, it precipitates as condensate. When a humid gas is cooled, dew forms, hence the term "dew point temperature". As soon as condensation occurs, the maximum saturation value for the water vapour in the gas is exceeded or the dew point is reached.

What appears here somewhat abstract is in practice one of the most everyday things in life. The effect is known, more or less consciously, to every spectacle wearer who enters a heated room in the cold season: the spectacles fog up, because the saturation value for water vapour in the air has been exceeded on the cold lenses.

Like absolute humidity, the dew point temperature is not pressure-dependent. However, a distinction is made between the terms "atmospheric dew point" and "pressure dew point". In compressed air networks the dew point temperature is an important variable for the quality of the compressed air. For the operator of compressed air systems, the pressure dew point is always decisive for the safe operation of the supply network. If a gas volume is compressed, the water vapour partial pressure increases, as already mentioned. This inevitably results in a higher dew point temperature.

In practice this means: If, metrologically speaking, a dew point of approximately −50 °C is obtained under atmospheric conditions, a pressure dew point of approximately −30 °C would result under an operating pressure of 10 bar. The relationship between atmospheric dew point and pressure dew point is shown in the diagram in Fig. 5.5. In this diagram different system pressures from 1 to 51 bar are given for the conversion from atmospheric dew point to pressure dew point and vice versa.

Figure 5.6 is used to convert the dew point temperature into ppmv (ppm(Vol)). A system pressure from 1013 mbar is selected as the reference value.

5.2.3 Relative Humidity in Gases

An important moisture parameter for determining the moisture content in gas flows is the relative humidity (r. h.) in percent. It provides information about the percentage saturation of gas with water vapor at a certain temperature. This measured variable is derived from

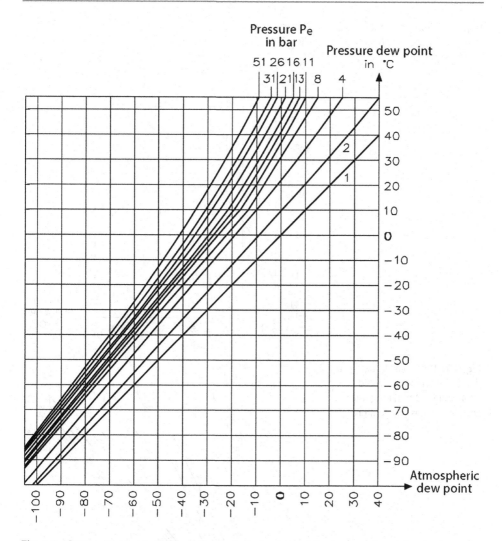

Fig. 5.5 Comparison between the atmospheric dew point and the pressure dew point

the ratio of the water vapour partial pressure p_w to the saturation vapour pressure p_s at a certain temperature

$$\text{r.h.} = \frac{p_w}{p_s} \cdot 100\%$$

The relative humidity is usually applied at atmospheric pressure conditions. Since the saturation vapour pressure is proportional to the gas temperature, it follows that it is strongly dependent on temperature. The relationship between dew temperature and relative humidity, taking into account the gas temperature, is shown in Fig. 5.7.

Fig. 5.6 Humidity concentration as a function of the dew point temperature

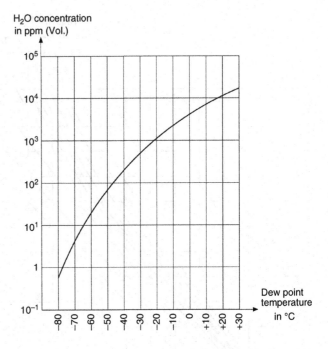

Fig. 5.7 Relationship between dew point temperature and relative humidity taking into account the gas temperature

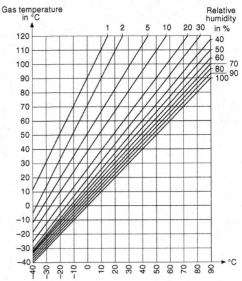

5.2.4 Relative Humidity in Liquids

As with moisture measurement in gases, the water content in non-aqueous liquids (oil, petrol, etc.) is often given in ppm, but here it is weight-related. Mathematically, the absolute humidity measured variable results from the following mathematical relationship:

Table 5.1 Liquids and their water saturation concentration in ppmw at a given temperature

Liquid	0 °C	10 °C	20 °C	30 °C	40 °C
Gasoline		454	639	870	1178
Heptane	27	54	96	172	308
Hexane			101	179	317
Octane		51	160	184	315

$$\text{ppmw} = [b]\frac{C_S}{p_S} \cdot p_W$$

C_S = Saturation concentration at a given temperature

p_S = Saturation vapor pressure at a given temperature

p_W = Water vapor partial pressure

The saturation concentration varies depending on the liquid and increases with increasing temperature. Table 5.1 lists various liquids with their water saturation concentration in ppmw at a given temperature.

5.2.5 Structure and Operation of an Alumina Humidity Sensor

Originally, the alumina humidity sensor was developed for research purposes in the upper regions of the atmosphere or during space flights. Since it proved itself under these extreme conditions, it was obvious to use it for the hardly less stringent requirements of process engineering. Figure 5.8 shows the design.

The critical point in manufacturing a sensor for measuring the partial pressure of water vapour is to maintain the correct thickness of the aluminum oxide film. Through continuous development it has been possible to create sensors that measure absolute values and are free of temperature and hysteresis effects. They respond quickly to fluctuating moisture concentrations and offer maximum stability. Thus, the water vapour partial pressure in gaseous and liquid media can be measured without problems.

The humidity sensor consists of an aluminium strip to which a wafer-thin oxide layer of defined thickness is applied by a special electrolytic process. A fine gold coating forms the outer electrode of the capacitor thus created between the aluminium and the gold layer.

Although electrically conductive, the gold coating is so permeable to water vapour that an equilibrium between the humidity of the surrounding medium and the water content of the aluminium oxide pores can be established within a short time. Figure 5.9 shows the equivalent electrical circuit of this sensor. The water vapour penetrates the gold layer and is deposited within the porous wall of the aluminium oxide. The number of water molecules absorbed here determines the conductivity and dielectric constant of the alumina. Thus, the electrical impedance of the arrangement changes approximately proportionally to the water content of the surrounding medium, regardless of whether it is a gas or a liquid. Figure 5.10 shows an alumina humidity sensor for certain gases and liquids.

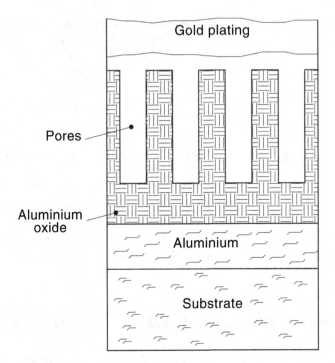

Fig. 5.8 Design of an alumina humidity sensor

Fig. 5.9 Cross-section and an electrical equivalent circuit of an aluminium oxide sensor with gold coating

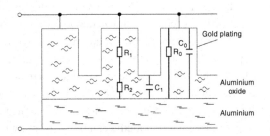

Because the pores of the aluminium sensor are so small that much larger molecules than water cannot penetrate, the water content of organic liquids can be measured in the same way as the water vapour partial pressure in gases.

5.2.6 Use of Alumina Humidity Sensors

Humidity sensors are suitable for determining the dew point over a range from −110 to +60 °C. This corresponds to a moisture content of approximately 0.001 to 200,000 ppmv (particles per million by volume) under atmospheric conditions. In principle, this measuring range also applies to liquids, but due to the process, it can never be fully exploited.

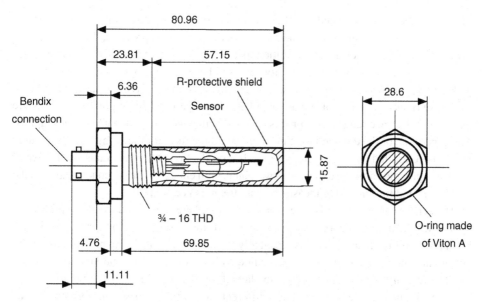

Fig. 5.10 Aluminium oxide humidity sensor for gases and liquids

Fig. 5.11 Measuring sensor in inline and measuring cell application

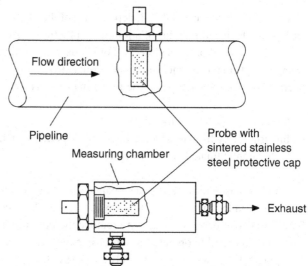

The sensors are installed directly in the process stream, exactly where the measurement is required (Fig. 5.11). For other systems with comparable characteristics, a partial flow must usually be taken. The need for piping, branches, flow, and pressure controllers is therefore reduced to a minimum. The ability to measure directly in the medium is particularly important for low humidity values and rapid humidity changes.

Any measuring system that requires a possibly exactly defined gas flow suffers from the influences of a bypass measurement, otherwise the accuracy and response speed would be ideal since the response speed and thus the instantaneous measured value depends on the respective moisture equilibrium state reached in the pipes. Gas pressure, flow rate, materials used, length and a clear width of the pipeline, etc. are disturbing factors which influence the time response and the humidity values to be tested in a bypass measurement.

Since it is not necessary to feed the process stream to the analyzer, dead times and possible disturbances due to leakage are avoided. This is particularly important for moisture analyses in the trace range or for process processes in which the moisture changes rapidly. The generation of the defined moisture content and the further processing of the measured values is controlled by a computer system. If a production stream contains conductive or corrosive particles or if the expected dew point is high enough to cause condensation in the system, a conditioning of the measuring gas stream is necessary.

The sensor can therefore be used on-site, that is, here is no need to take samples and then bring them to the laboratory for analysis. In the descriptions of various analytical methods for moisture analysis, it is repeatedly pointed out that the main problem lies in the change in water concentration in organic liquids between sampling and laboratory analysis. The reason for this can be easily explained by "Henry's" law. If the volume of gas above the liquid is more humid or has a different temperature, a new equilibrium between the gas and the liquid is established, which inevitably changes the water content in the liquid.

Due to the pressure and flow independence of the alumina humidity sensor, it is particularly suitable for direct installation in a pipeline system, such as in compressed air networks or closed process systems. For the user, this ultimately means rapid detection of changes and unaltered measurement results. However, sometimes it is unavoidable to operate the sensor in a bypass. The following reasons can be decisive for this:

- Medium temperatures too high
- Impermissibly high flow velocities in the pipeline system
- highly contaminated gases or liquids

Another reason to work with a so-called sampling system is the fact that if the humidity sensor is installed directly in a closed system, it may be necessary to shut down the process flow partially or even completely to remove the humidity sensor. In this case, it is in any case recommended to operate the sensor in a bypass system.

Figure 5.12 shows such a relatively simple system; it can be realized with a connection piece in the mainline, a needle valve, a measuring chamber to accommodate the humidity sensor, and a stainless steel spiral to prevent back diffusion.

Depending on the measuring task, such a sampling system can also be somewhat more complex. Sampling systems are generally suitable for

- reducing impermissibly high medium temperatures
- to filter out solid particles
- regulate flow rates

Fig. 5.12 Humidity sensor in a bypass system

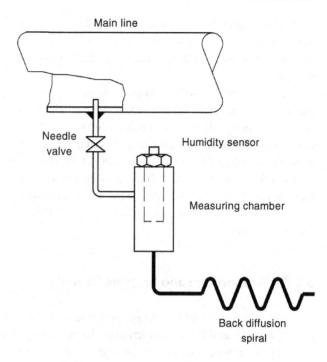

* reduce pressures
* the suction of measuring media in unpressurized systems
* to combine several measuring points centrally

Even the smallest leaks in a system in which the lowest humidity contents or low dew point temperatures are to be measured lead to considerable falsification of the measured values. It is essential to ensure that all screw connections used, especially compression fittings, are properly tightened by the manufacturer's specifications. Furthermore, it is important to ensure that rubber or plastic hoses are not used as supply material, especially in low dew point areas. Although some high-quality plastics give the impression that no moisture can penetrate the hoses, water molecules still diffuse through the hose wall into the interior and thus falsify the measurement result. To guarantee a maximum of measuring reliability, all parts in contact with the medium must be made of stainless steel, such as needle valves, piping, measuring chambers, filters, etc. If the humidity sensor is operated in bypass and the gas to be measured is discharged against ambient air, it is essential to install a spiral to prevent back diffusion into the outlet of the measuring section.

The response time of the sensor to changes in humidity is specified as 7 s for 63% of the final value. Of course, this time can only be achieved if the sensor is located directly in the pipe or, when using a bypass system, the distance between the sampling point (dead space) and the sensor is kept as short as possible. It should also be noted that adsorption and desorption processes on the supply line walls etc. can have a strong negative effect on the response speed of the humidity sensor. The response speed is equally dependent on the flow velocity, that is, if a humid system is exposed to a dry gas or liquid, the setting of the final state is reduced with an increase in the flow velocity, as the system is flushed dry more quickly.

The sensor can be used in a large, dynamic flow rate range. Thus, moisture contents in gases can be measured at flow rates of static conditions up to 10 m/s under atmospheric pressure, or in liquids at static conditions up to 10 m/s, at a density of 1 g/m^3. The sensor has no flow dependence, but higher flow velocities favour a faster drying out of the humid system and thus enable a faster adjustment of the final moisture value.

The often observed effect that the measured value changes when the flow velocity increases are not due to a flow dependence, but rather water molecules adhering to the pipe walls are torn off and flushed along more quickly, thus giving the impression of a higher measured value.

5.2.7 Temperature and Pressure Behavior

The aluminium oxide sensor can be used at temperatures up to +70 °C. Higher temperatures inevitably lead to the destruction of the measuring element. Because the adsorption tendency of the water molecules is strongly temperature-dependent, the undesired adsorption effects are considerably reduced by heating the measuring section.

In general, the medium temperature should be kept largely constant, absolutely no changes in the humidity profile should be obtained due to the temperature-dependent adsorption behavior of the water molecules. This effect occurs in the case of outdoor lines in day-night operation, that is, with constant humidity conditions, the humidity content changes with temperature, but this is exclusively system-related.

If it is necessary to examine gases with very high temperatures for their moisture content, the sample gas must be cooled before the sensor. For this purpose, it is usually sufficient to install a cooling spiral in front of the sensor. This allows gas temperatures of approximately 2000 °C to be reduced to room temperature; however, care must be taken to ensure that the temperature does not fall below the dew point during cooling.

A dew point shortfall occurs as soon as the difference between dew point temperature and medium temperature becomes too small and condensation occurs in the system. To avoid this unwanted effect, a minimum difference of approximately 10 °C between the dew point temperature and the medium temperature should be maintained.

Depending on the mechanical design, this sensor can be used up to pressures of 350 bar, but it must be ensured that it is not suddenly exposed to higher pressures. The pressure should be built up or released as slowly as possible to protect the sensor from mechanical destruction by pressure waves. When measuring the dew point, it must be ensured that the measurement result reflects the pressure dew point, that is, the actual dew point temperature at the existing system pressure.

5.3   Realization of Humidity Measurement

There is a wide range of interest in the measurement of air humidity. The weather report always mentions relative humidity. This indicates the percentage related to the saturation humidity. Example: There is an air temperature of 25 °C and the air can absorb approximately 25 g water per m³. If you measure 15 g/m³ absolutely, then at this temperature 60% corresponds to the possible maximum, i.e. a relative humidity of 60%.

In addition to the purely informational measurement of humidity, as it is done for example in many private households, offices, and business premises, there is the measurement of an exact value, which is required in automated systems that automatically produce a certain humidity and thus maintain a control system. In the first case, the well-known hair hygrometer is often used, which works purely mechanically and is not very accurate. In the second case, a precise humidity sensor is required, which gives a signal as proportional as possible to the respective humidity. There are several different principles for measuring the humidity of the air. Correspondingly different are also the sensors and the effort involved.

5.3.1   Simple Measuring Circuit with Humidity Sensor

The humidity sensor used here consists of a special foil vaporized with a gold film on both sides. This represents the dielectric of a plate capacitor, the two gold films form its electrodes. Under the influence of the air humidity, the dielectric constant of the film changes and thus the capacity of the capacitor. With the help of a simple measuring circuit, the capacitance or its change is recorded and converted into a DC voltage. This can then be used for direct display of the relative humidity or as an actual value for a system for automatic humidity control.

The sensor element is installed in a perforated plastic housing (Fig. 5.13), which is suitable for direct mounting on a circuit board.

Fig. 5.13 Dimensions of the humidity sensor

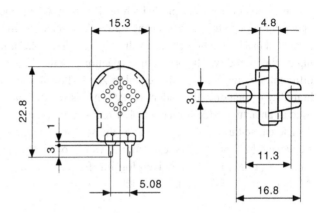

Fig. 5.14 Capacity C_s of the sensor as a function of relative humidity

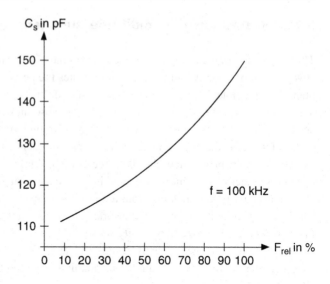

Fig. 5.15 Principle circuit of a measuring bridge for recording the changes in capacitance of the humidity sensor

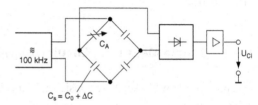

Figure 5.14 shows the dependence of the capacity C_s of the sensor on the relative humidity, it is not linear. A directly measured value display, therefore, requires either a measuring instrument with an appropriately adjusted scale or a linearization of the display using circuitry measures.

The conversion of the capacity change into a corresponding electrical signal is possible according to different principles. In practice, a measuring bridge as shown in Fig. 5.15 is often used.

An oscillator, for example, which oscillates at a frequency of 100 kHz, generates the operating voltage for the measuring bridge. The sensor with the capacitance $C_s = C_0 + \Delta C$ is located in a bridge branch. The value C_0 is the fixed, ΔC is the capacitance share depending on the humidity. The balancing capacitor C_A must be set so that for $\Delta C = 0$ the bridge differential voltage ΔU is also equal to zero. The value for ΔU which is set for a certain humidity is then only dependent on ΔC. ΔU is rectified and amplified, then displayed or used as an actual value for a control or regulation system.

The capacitance $C_s = C_0 + \Delta C$ of the sensor is slightly dependent on the measuring frequency. In Table 5.2, the capacitance values of C_s at $F_{rel} = 0\%$ (here referred to as C_0) and for C_s at $F_{rel} = 12\%$ are given for four frequencies, as well as the capacity difference ΔC between $F_{rel} = 0\%$ and $F_{rel} = 100\%$.

Table 5.2 Capacitance values of the humidity sensor for different frequencies

Frequency	C_0 ($F_{rel} = 0\%$)	C_s ($F_{rel} = 12\%$)	ΔC ($F_{rel} = 0 \ldots 100\%$)
1 kHz	116.1 pF	119.7 pF	45.5 pF
10 kHz	112.7 pF	116.2 pF	44.2 pF
100 kHz	109.0 pF	112.3 pF	42.7 pF
1 MHz	104.6 pF	107.9 pF	41.0 pF

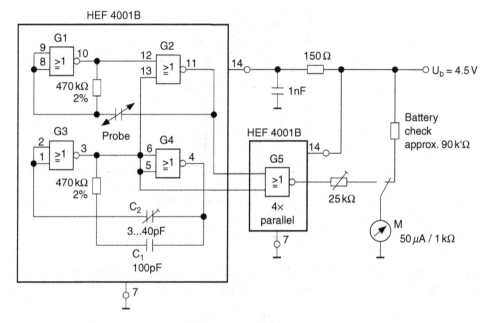

Fig. 5.16 Simple measuring circuit for moisture detection

The figures in the table are typical values; due to manufacturing variations, they may vary within the permitted tolerances. However, the relative change in these values due to frequency is practically equal to the relative changes in typical values.

Figure 5.16 shows a measurement circuit with five NOR gates, contained in two CMOS devices of type 4001. The circuit works on the principle of differential pulse measurement. The range of the operating voltage can be between 4.5 and 9 V, the current consumption is around 100 pA at an average humidity.

The four NOR gates $G_1, \ldots, G_4$ form two rectangular generators M_1 and M_2. Square wave generator M_1 oscillates freely with a frequency determined by the 470-kΩ resistor and the total capacitance of the two capacitors C_1 and C_2. In this example, a frequency of 10 kHz is generated. Square wave generator M_2 is synchronized by M_1, so it operates at the same frequency as M_1. The pulse duration of Fig. 5.17 depends on the capacitance $C_0 + \Delta C$ of the sensor and thus on the humidity. Between the two square-wave generators, which are linked via NOR gate G5, differential pulses with the duration $t_3 = t_2 - t_1 \approx \Delta C$ occur if the same proportionality factor applies to both. If, for example, the period T of the

Fig. 5.17 Pulse diagram of difference formation for the circuit of Fig. 5.16

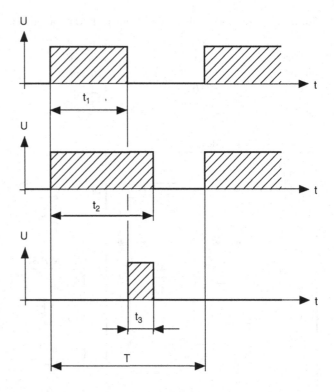

square-wave voltages is selected as $T = 2 \cdot t_1$ and all pulses use the amplitude, then the following applies to the arithmetic mean value of the output voltage U_0

$$U_{0(\mathrm{AV})} = \frac{t_3}{T} \cdot U_b = \frac{\Delta C}{2 \cdot C_0} \cdot U_b$$

The temperature and voltage dependence of the ratio t_3/T is very low if

- the properties of both rectangle generators are largely identical, e.g. when using the CMOS device 4001
- the sensor capacitance $(C_0 + \Delta C)$ and the trimming capacitor (C_2) have the same temperature coefficient

The pulses generated in the two square-wave generators are fed to the inputs of another CMOS component, which is controlled depending on the difference between the pulses. The four NOR gates of this second component are connected in parallel so that a relatively low impedance output is obtained, to which the display instrument (50 µA, 1 kΩ) is connected via a potentiometer from 25 kΩ.

Table 5.3 Scaling of the humidity sensor in the frequency range from 10 kHz to 1 MHz

F_{rel} [%]	0	10	20	30	40	50	60	70	80	90	100
Indication [%]	0	6.6	13.2	20.5	29.0	36.8	46.0	56.6	67.6	81.6	100

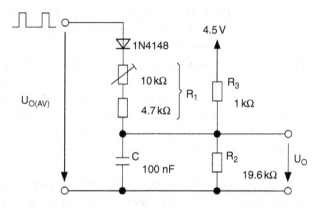

Fig. 5.18 Linearization circuit for the humidity sensor

Due to the non-linear relationship between sensor capacity and relative humidity, an indicating instrument with special graduation is required. For the frequency range from 10 kHz to 1 MHz the scaling table in Table 5.3 applies.

By additional circuitry measures, it is possible to achieve an extensive linearization of the display and to use an instrument with linear scale division. Figure 5.18 shows the linearization circuit which is connected to the output of the NOR gate.

The circuit of Fig. 5.18 is available in two versions, namely

- for the operation of a display instrument (50 µA, 1 kΩ)
- with voltage output (0 ... 1 V)

In principle, the linearization circuit works according to the following principle: The output pulses formed in the measuring circuit charge capacitor C via diode D (1N4148) and resistor R_1, while at the same time a discharge current proportional to the capacitor voltage flows via R_2. The output voltage occurring at the capacitor is not proportional to U_0' the arithmetic mean value of the pulse output voltage $U_{O(AV)}$. By suitable dimensioning of the components C, R_1, and R_2 in conjunction with an additional current via R_3, the function can be $U_0' = f(U_0)$ designed in such a way that the relationship between the output voltage U_0' and the measured quantity F has a substantially improved linearity.

The operating voltage goes directly into the measured value $U_{O(AV)}$; it may therefore have to be stabilized. The simple circuit can be constructed at very low cost, it works reliably and with an accuracy sufficient for many applications. When supplied by three mignon cells, continuous operation of about 1 year is possible.

The circuit is calibrated at an operating voltage of +4.5 V; instead of the humidity sensor, a capacitor of 118 pF is used for the basic calibration. Then C_2 is adjusted to the minimum display or output voltage. Then the 118 pF capacitor is replaced by one with 159 pF, and the circuit is adjusted with the 10 or 25 kΩ dial to a full scale of the pointer instrument or to an output voltage of 1 V. Only now can the humidity sensor be installed. Afterward, C_2 is used for calibration until the setpoint value of the existing humidity (recommended for $F_{rel} \approx 50\%$) is displayed or the corresponding output voltage appears.

If the circuit is calibrated at an average humidity, the highest measuring accuracy is obtained in this important range, but the possible error is greater at the ends of the range. Based on the measurement results given, the measurement error under the most unfavorable conditions should be about 5% for the stabilized circuit in the middle of the range and about 8% for 10% and 90% humidity. With an unstabilized operating voltage, the values are assumed to be doubled. Under normal conditions, that is, at room temperature and the nominal operating temperature, which is present at the extraordinarily low current consumption for the largest part of the operating time, the display errors should be well below these limits.

Saturated salt solutions are suitable for adjusting or calibrating the humidity sensor circuit. Here one uses the fact (see DIN 40046) that the relative humidity of the air, which adjusts itself in a closed container over a saturated salt solution contained in it, has a certain value that depends only on the temperature. The sensor is arranged in an airtight container in such a way that its connecting pins are guided through the container wall and can be contacted with the circuit from outside. Then a cotton ball soaked with the saturated solution is placed in the container and sealed airtight. After a waiting period of at least 30 min at a constant temperature, the circuit can be calibrated.

5.3.2 Humidity Dependent Control

Humidity dependent control often requires a circuit that triggers a switching operation at a given relative humidity. If the relative humidity measured by the humidity sensor is exceeded, a relay picks up. If the relative humidity falls below this value, the relay drops out again. The switching hysteresis is adjustable. Two LEDs for the operating voltage and the relay signal the operating conditions.

The circuit of Fig. 5.19 can be used as a detector, signaling, or warning device for condensation water and humidity. The response threshold is continuously adjustable. The circuit is not intended to indicate humidity. The sensor has a certain response time. In case of sudden changes in humidity (e.g. during transport to another environment), it is, therefore, necessary to wait a few minutes until the measuring signal follows.

At a humidity of 0%, the sensor has a capacity of approximately 105 ... 115 pF, at a humidity of 105% approximately 145 ... 160 pF. The variation range is thus approximately 40 ... 45 pF or approximately 40% relative to the basic capacity.

Such variations can easily be captured indirectly. To do so, you build up an oscillator with the variable capacitor and detect its frequency changes. The circuit of Fig. 5.19 works as follows: The RC-oscillator with three CMOS-NON-gates delivers a square wave

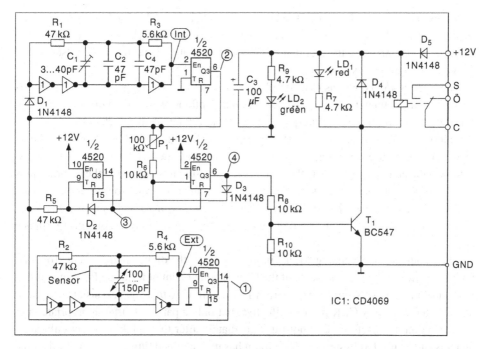

Fig. 5.19 Humidity dependent control with threshold value behaviour

frequency of about 300 kHz as a reference signal. This output is marked "Int". The down-stream four-stage binary counter divides this frequency by 16, so that at point 2 there is approximately 20 kHz, corresponding to a period of 50 μs.

The identical RC oscillator at the bottom left uses the variable sensor capacitance as the frequency-determining element. Here the square wave signal "Ext" is generated, which also oscillates with approximately 300 kHz and at point 1 again with approximately 20 kHz. The Ext oscillator (and point 1) runs permanently and also clocks the 4-bit counter IC3.1 (CMOS component 4520). This counter locks itself and its neighbour IC3.2 (4520), if after eight CLK-pulses the output Q3 (point 3) switches to 1-signal. If this is the case, point 4 is continuously at 0 signal and the transistor with its relay blocks.

By setting point 3, the Int oscillator can prevent the Ext output from oscillating at a higher frequency. If this is the case, it resets counter 4520 to 0 just before the eighth pulse is reached, by a 1 signal from point 2, which is connected to RESET. At the same time, point 2 clock counts up the counter 4520 (via P_1 and R_6), and after the eighth CLK pulse, point 4 goes to 1 signal. Diode D3 now blocks its clock input and now the transistor receives the necessary base current via resistor R_8. This is amplified and switches the relay on. The case occurs when the Int-frequency is higher than the Ext-frequency.

As already mentioned, the capacity of the sensor increases with increasing humidity, so that the oscillator frequency decreases with it. If the humidity exceeds a certain value, the relay switches on. The upper oscillator can only oscillate during the 0-signal times of point

1, because it is blocked by diode D_1 when the 1-signal is at 1. This establishes the defined oscillation and the synchronisation between both oscillators. The eight pulses on the 16-divider are caused by the fact that the last output is at 1-signal for eight pulses and at 0-signal for eight pulses.

The CMOS 4520 contains two complete 4-bit counters. The respective clock signal is applied via the T input (pins 1 and 9). The internal flipflops operate with the positive clock edge at the T input. If the E_n input (pins 2 and 10) has a 1 signal, the clock pulses can reach the internal counter. If the signal is 0, the respective T input is disabled. However, the E_n input can also be used as a frequency input, in which case the frequency input can be disabled via the actual T input, as is the case here. If a 1 signal is applied to the R input (pins 7 and 15), the internal flipflops of the respective counter are reset to 0, that is, the four outputs of a counter then have a 0 signal. The outputs Q_0 (pins 3 and 11), Q_1 (pins 4 and 12), and Q_2 (pins 5 and 13) are not required in this application, only the outputs Q_0 (pins 6 and 14).

The P_1 adjuster is an analog "trick" in the middle of the digital environment. Together with the parasitic input capacitance of the CMOS component, it forms an RC element that loops and attenuates the edges at CLK (pin 1). Therefore, with a higher resistance value of the adjuster, not every CLK pulse is effective, but only a part of it, after appropriate integration. Point 4, therefore, does not tilt immediately after the eighth beat, but only later, and this can be used as hysteresis. The circuit has no undefined tilting back and forth at the moment of switching.

Once the circuit has been set up, connect the humidity sensor to the pins marked F-Sensor. A longer cable connection between the sensor and measuring circuit should be avoided if possible, as the circuit capacity increases and leads to a reduction of the duty cycle. If a cable must be inserted between the sensor and the circuit board, a reference capacitance must be inserted.

For adjustment, turn the slider of the adjuster P_1 to the left stop. If the operating voltage is now switched on, the green LED must emit light. The variable capacitor C_1 may only be adjusted with a plastic screwdriver since a metallic one would strongly falsify the capacity. Adjust the capacitor C_1 until the relay drops out, then bring the circuit into an environment with the humidity you want to measure. After a waiting period of 3–5 min, the switching point is adjusted by turning the variable capacitor C_1, and then the adjuster is moved to the middle position. If the response behavior at the moment of changeover is too unstable or too slow, this can be corrected with the adjuster.

A very precise reference point can be established by using a sealed container containing a saturated sodium chloride solution (common salt). The air in this container constantly assumes a relative humidity of 75% in the range from 25 to 50 °C.

Index

© Springer Fachmedien Wiesbaden GmbH, part of Springer Nature 2022
H. Bernstein, *Measuring Electronics and Sensors*,
https://doi.org/10.1007/978-3-658-35067-3

Printed in the United States
by Baker & Taylor Publisher Services